Mathematical Statistical and Financial Tables for the Social Sciences

Z.W. Kmietowicz
and Y. Yannoulis

Longman
London and New York

Longman Group Limited London

Associated companies, branches and representatives throughout the world

Published in the United States of America by Longman Inc., New York

First published 1976

Library of Congress Cataloging in Publication Data

Kmietowicz, Z. W.
Mathematical, statistical, and financial tables for the social sciences.

1. Mathematics – Tables, etc. 2. Mathematical statistics – Tables, etc. 3. Interest and usury – Tables, etc. I. Yannoulis, Y., joint author. II. Title.

QA47.K58 510'.21'2 75–42473
ISBN 0 582 44062 9

Set in IBM Journal
and printed in Great Britain by
Whitstable Litho Ltd, Kent

Useful data

$e = 2.71828$; $e^2 = 7.38906$; $1/e = 0.36788$; $\sqrt{e} = 1.64872$

$\pi = 3.14159$; $\pi^2 = 9.8696$; $1/\pi = 0.31831$; $\sqrt{\pi} = 1.77245$

$\sqrt{10} = 3.16228$; $\sqrt[3]{10} = 2.15443$; $\sqrt[3]{100} = 4.64159$; $\sqrt{2} = 1.41473$

$\log_{10}e = 0.43429$; $\log_e 10 = 2.30259$; $\log_{10}\pi = 0.49715$; $\log_e \pi = 1.14473$; $\log_e x = \log_e 10 \,.\, \log_{10} x$; $\log_{10} x = \log_{10} e \,.\, \log_e x$; $\log_e 10 = 1/\log_{10} e$.

Area under Normal curve between: $\mu \pm \sigma = 0.68268$; $\mu \pm 2\sigma = 0.95450$; $\mu \pm 3\sigma = 0.99730$; $\mu \pm 1.6449\sigma = 0.9$; $\mu \pm 1.9600\sigma = 0.95$; $\mu \pm 2.5758\sigma = 0.99$.

Circle: $\pi = \frac{C}{D} = \frac{C}{2r}$ (r = radius; D = diameter; C = circumference), $C = D\pi = 2r\pi$.

Area of: square = l^2 (l = length of one side); rectangle = base × height; parallelogram = base × perpendicular height; triangle = $\frac{1}{2}$ base × height; circle = πr^2; sphere = $4\pi r^2$; curved surface of cylinder = $2\pi rh$ (h = height); trapezium = height × average length of parallel sides.

Volume of: cube = l^3; sphere = $\frac{4}{3}\pi r^3$; cylinder = $\pi r^2 h$; cone = $\frac{1}{3}\pi r^2 h$; pyramid = $\frac{1}{3}$ area of base × height.

Metric measures

Length

1000 micrometres	= 1 millimetre (mm)
10 millimetres	= 1 centimetre (cm)
10 centimetres	= 1 decimetre (dm)
10 decimetres	= 1 metre (m)
10 metres	= 1 decametre (dam)
10 decametres	= 1 hectometre (hm)
10 hectometres	= 1 kilometre (km)

Weight

1000 microgrammes	= 1 milligramme (mg)
10 milligrammes	= 1 centigramme (cg)
10 centigrammes	= 1 decigramme (dg)
10 decigrammes	= 1 gramme (g)
10 grammes	= 1 decagramme (dag)
10 decagrammes	= 1 hectogramme (hg)
10 hectogrammes	= 1 kilogramme (kg)
1000 kilogrammes	= 1 tonne

Volume and capacity

10 millilitres (ml)	= 1 centilitre (cl)
10 centilitres	= 1 decilitre (dl)
10 decilitres	= 1 litre (l)
10 litres	= 1 decalitre (dal)
10 decalitres	= 1 hectolitre (hl)
10 hectolitres	= 1 kilolitre (kl)

Area

1 are	= 100 square metres (m^2)
1 decare	= 10 ares
1 hectare	= 100 ares
1 deciare	= 1/10 of an are
1 centiare	= 1/100 of an are
	= 1 square metre (m^2)

Conversion (metric to British and British to metric)

Length

1 cm	= 0.39370 in
1 m	= 39.37011 in
	= 3.28084 ft
	= 1.09361 yd
1 km	= 0.62137 mile
1 in	= 2.54000 cm
1 yd	= 0.91440 m
1 mile	= 1.60934 km

Weight

1 g	= 0.03527 oz
1 kg	= 2.20462 lb
1 tonne	= 0.98421 ton
1 oz	= 28.34953 g
1 lb	= 0.45359 kg
1 ton	= 1.01605 tonne

Volume and capacity

1 cubic cm	= 0.06102 cubic in (in^3)
1 cubic m	= 1.30795 cubic yd (yd^3)
1 litre	= 1.75980 pints
	= 0.21998 gallon
1 cubic in	= 16.38702 cubic cm (cm^3)
1 cubic yd	= 0.76455 cubic m (m^3)
1 pint	= 0.56825 litre
1 gallon	= 4.54596 litres

Area

1 cm^2	= 0.15500 in^2	1 in^2	= 6.45159 cm^2
1 m^2	= 1.19599 yd^2	1 yd^2	= 0.83613 m^2
1 km^2	= 0.38610 m^2	1 m^2	2.58998 km^2
1 hectare (ha)	= 2.47106 acres	1 acre	= 0.40469 hectare (ha)

Basic data about

probability distributions

Distribution	$f(x)$	$\bar{x} = E(x) = \mu'_1$	$\sigma^2 = \text{var}(x) = \mu_2$	Moment coefficient of skewness $(\mu_3/\mu_2^{3/2})$	Moment generating function
Discrete					
1. Uniform	$\frac{1}{n}$ $x = k + h, k + 2h, \ldots k + nh$	$k + \frac{1}{2}h(n + 1)$	$\frac{1}{12}h^2(n^2 - 1)$	0	$\frac{e^{(k+h)t}(1 - e^{nht})}{n(1 - e^{ht})}$
2. Binomial $(p + q = 1)$	${}^nC_x p^x q^{n-x}$ $x = 0, 1, 2, \ldots n$	np	npq	$(q - p)/\sqrt{npq}$	$(q + pe^t)^n$
3. Poisson	$e^{-m}\frac{m^x}{x!}$ $x = 0, 1, 2, \ldots$	m	m	$1/\sqrt{m}$	$\exp[m(e^t - 1)]$
4. Geometric $(p + q = 1)$	pq^{x-1} $x = 1, 2, 3, \ldots$	$\frac{1}{p}$	$\frac{q}{p^2}$	$\frac{1 + q}{\sqrt{q}}$	$\frac{pe^t}{1 - qe^t}$
5. Negative Binomial $(p + q = 1)$	${}^{r+x-1}C_x p^r q^x$ $x = 0, 1, 2, \ldots$	$\frac{rq}{p}$	$\frac{rq}{p^2}$	$\frac{1 + q}{\sqrt{rq}}$	$\left(\frac{p}{1 - qe^t}\right)^r$
Continuous					
6. Rectangular (Uniform)	$\frac{1}{b - a}$ $a \leqslant x \leqslant b$	$\frac{1}{2}(a + b)$	$\frac{(b - a)^2}{12}$	0	$\frac{e^{bt} - e^{at}}{t(b - a)}$
7. Negative Exponential	$\theta e^{-\theta x}$ $x > 0$	$\frac{1}{\theta}$	$\frac{1}{\theta^2}$	2	$\frac{\theta}{\theta - t}$
8. Normal	$\frac{1}{\sigma\sqrt{2\pi}}\exp\left\{-\frac{1}{2}\left(\frac{x - \mu}{\sigma}\right)^2\right\}$ $-\infty < x < +\infty$	μ	σ^2	0	$\exp[\mu t + \frac{1}{2}\sigma^2 t^2]$
9. Gamma $\alpha > 0, \beta > 0$	$\frac{1}{\beta^\alpha \Gamma(\alpha)} x^{\alpha - 1} e^{-x\beta}$ $x > 0$	$\alpha\beta$	$\alpha\beta^2$	$2/\sqrt{\alpha}$	$(1 - \beta t)^{-\alpha}$
10. χ^2 (Chi-squared)	$\frac{1}{2^{\nu/2}\Gamma(\frac{\nu}{2})} x^{(\frac{\nu}{2} - 1)} e^{-x/2}$ $x > 0$	ν	2ν	$2\sqrt{\frac{2}{\nu}}$	$(1 - 2t)^{-\nu/2}$

Basic data about probability distributions – *continued*

Distribution	$f(x)$	$\bar{x} = E(x) = \mu_1'$	$\sigma^2 = \text{var}(x) = \mu_2$	Moment coefficient of skewness $(\mu_3/\mu_2^{3/2})$
Discrete 11. Hypergeometric	$\dfrac{{}^{Np}C_x \, {}^{Nq}C_{n-x}}{{}^{N}C_n}$ $x = 0, 1, 2, \ldots n$	np	$npq\left(\dfrac{N-n}{N-1}\right)$	$\dfrac{(q-p)(N-2n)\sqrt{N-1}}{(N-2)\sqrt{npq}(N-n)}$
Continuous 12. Pareto $\alpha > 0, \beta > 0$	$\alpha\beta^{\alpha}x^{-(\alpha+1)}$ $x > \beta$	$\dfrac{\alpha\beta}{\alpha - 1}$	$\dfrac{\alpha\beta^2}{(\alpha-2)(\alpha-1)^2}$	$\dfrac{2(\alpha+1)\sqrt{\alpha-2}}{\sqrt{\alpha}(\alpha-3)}$
13. Lognormal	$\dfrac{1}{x\sigma\sqrt{2\pi}}\exp\left\{-\dfrac{1}{2\sigma^2}(\log_e x - \mu)^2\right\}$ $x > 0$	$e^{\mu + \frac{1}{2}\sigma^2}$	$e^{2\mu+\sigma^2}(e^{\sigma^2} - 1)$	$(e^{3\sigma^2} - 1)/(e^{\sigma^2} - 1)^{3/2}$
14. Beta $\alpha > 0, \beta > 0$	$\dfrac{\Gamma(\alpha+\beta)}{\Gamma(\alpha)\Gamma(\beta)}x^{\alpha-1}(1-x)^{\beta-1}$ $0 < x < 1$	$\dfrac{\alpha}{\alpha+\beta}$	$\dfrac{\alpha\beta}{(\alpha+\beta)^2(\alpha+\beta+1)}$	$\dfrac{2(\beta-\alpha)\sqrt{\alpha+\beta+1}}{(\alpha+\beta+2)\sqrt{\alpha\beta}}$
15. t	$\dfrac{\Gamma\left(\frac{\nu+1}{2}\right)\left(1+\frac{x^2}{\nu}\right)^{-\left(\frac{\nu+1}{2}\right)}}{\sqrt{\nu\pi}\,\Gamma(\nu/2)}$ $-\infty < x < +\infty$	0	$\dfrac{\nu}{\nu-2}$ $\nu > 2$	0
16. F	$\dfrac{\Gamma\left(\frac{\nu_1+\nu_2}{2}\right)\nu_1^{\frac{\nu_1}{2}}\nu_2^{\frac{\nu_2}{2}}x^{\left(\frac{\nu_1-2}{2}\right)}}{\Gamma\left(\frac{\nu_1}{2}\right)\Gamma\left(\frac{\nu_2}{2}\right)(\nu_2+\nu_1 x)^{\left(\frac{\nu_1+\nu_2}{2}\right)}}$ $x > 0$	$\dfrac{\nu_2}{\nu_2 - 2}$ $\nu_2 > 2$	$\dfrac{2\nu_2^2(\nu_1+\nu_2-2)}{\nu_1(\nu_2-2)^2(\nu_2-4)}$ $\nu_2 > 4$	$\dfrac{2(2\nu_1+\nu_2-2)\sqrt{2(\nu_2-4)}}{(\nu_2-6)\sqrt{\nu_1(\nu_1+\nu_2-2)}}$ $\nu_2 > 6$

Contents

Acknowledgements

During the preparation of this publication we obtained assistance from a number of colleagues engaged in teaching quantitative methods to Social Science students and we would like to mention their contribution.

Our greatest debt is to D. Harkess, Research Officer, School of Economic Studies University of Leeds, and to A. J. MacLeod, Research Officer, Department of Economics Queen's University, Belfast, for invaluable help with computing problems arising from compilation of most of the tables presented in this booklet. We would also like to thank for their helpful comments and suggestions concerning the contents Professor C. E. V Leser, A. D. Pearman, E. G. Drettakis, R. A. Hart, A. G. Witts, F. W. A. Zanker and D. G Slattery (Economics), E. Graham (Business Studies), A. B. Royse (Sociology), J. Hartley (Education) and H. McAllister (Psychology). We also want to express our thanks to Professor D. F. Kerridge for permission to use his computer programs of the *t* and *F* distributions, and Professor K. V. Mardia and P. J. Zemroch for letting us use their computer program to obtain values of *F*-distribution for large numbers of degrees of freedom. Finally, we would like to thank the secretarial staff of our departments for their help with the preparation and typing of the manuscript, particularly Mrs M. M. Seddon Mrs T. E. Brier, Mrs A. Williams and Mrs M. Gilmore.

Publisher's acknowledgements

We are grateful to the following for permission to reproduce copyright material:

American Cyanamid Company for a table from *Some Rapid Approximate Statistical Procedures*, 1949, by F. Wilcoxon; American Psychological Association for a table from *Psychological Bulletin*, Vol. 51, 1954, by L. A. Goodman; American Statistical Association for a table by L. H. Miller from *Journal of American Statistical Association*, Vol. 51, 1956, and a table by W. H. Kruskal and W. A. Wallis from *Journal of American Statistical Association*, Vol. 47, 1952; Biometrika for an extract from a table in *Biometrika Tables for Statisticians*, Vol. 1, 1970; Imperial College of Science and Technology for values from tables by Durbin and Watson in *Biometrika*, Vol. 41, 1951; Holt, Rinehart and Winston Inc., for a table by H. Walker and J. Lev from *Statistical Inferences*, 1953; The Institute of Mathematical Statistics for tables from the *Annals of Mathematical Statistics*, Vol. 18 and Vol. 22 by H. B. Mann and D. R. Whitney, and the Indiana University for tables from the *Bulletin of the Institute of Educational Research*, Vol. 1, no. 2, by D. Aube, 1953.

Preface

The aim of this publication is to present under one cover a set of mathematical, statistical and financial tables widely used in social sciences, i.e. economics, management, business studies, sociology, politics, psychology, geography and education. At present the tables are scattered over a number of books and articles, and are not always in the form required by the general user. Here they are brought together in a convenient form and it is hoped that they will prove a useful reference.

The tables are primarily designed for teaching and research purposes in universities, polytechnics, colleges, and schools, but social scientists working outside the field of education should find them equally useful. Our experience indicates that a handy set of tables would be particularly useful to students and teachers of quantitative methods in classes and tutorials. The tables are also ideally suited for examination purposes as they provide all the tables likely to be needed.

It is assumed that the users are essentially familiar with the tables, nevertheless diagrams, short explanatory notes, and illustrative examples are provided to make the reference easier.

The tables are arranged in three sections. Section 1 contains commonly used mathematical tables. Tables 1 to 4, include logarithms, squares, square roots, reciprocals, etc., and are particularly useful for computational purposes. Tables 5, 6 and 7, include factorials, logarithms of factorials, and binomial coefficients, and serve a similar purpose for probability calculations.

Section 2 presents the most frequently used statistical tables and constitutes the core of the publication. Tables in this section are grouped for convenience into four subsections. Subsection (a) presents the basic probability distributions, i.e. Binomial, Poisson, Exponential, Normal, χ^2, t and F, on which many tests of significance are based. Subsection (b) contains tables needed for tests of significance in regression and correlation analysis, which is widely used in all social sciences. Tables 16, 17 and 18 are needed to test the significance of simple, partial and multiple correlation coefficients, and Table 19 is used to test for autocorrelation of regression residuals. Tables in Subsection (c) are used for nonparametric tests of significance, i.e. when ordinal, rather than cardinal, measurements are involved, and consequently are particularly relevant for sociology, politics, psychology and education. Tables of random numbers in Subsection (d) can be used to select random samples in survey investigations, to conduct statistical experiments, etc.

Section 3 contains the basic accounting tables used in financial decision making and management investment appraisal.

Z. W. Kmietowicz

Y. Yannoulis

Section 1

Mathematical tables

Notes on the use of Tables 1, 2 and 3

1. **Logarithms.** The logarithm of a positive number (x) consists of two parts, an integral part, called the characteristic, and the decimal part, called the mantissa, which is usually positive. For numbers greater than 1, the characteristic is one less than the number of digits to the left of the decimal point, e.g. characteristics of 858.3, 85.83 and 8.583 are 2, 1 and 0 respectively. For numbers between 0 and 1, the characteristic is negative, and is greater by one than the number of zeros that follow the decimal point, e.g. characteristics of 0.8583, 0.08583 and 0.008583 are $\bar{1}$, $\bar{2}$ and $\bar{3}$ respectively. To find the logarithm of a number consisting of four figures, e.g. log 858.3 find the first two numbers (85) in the first column of Table 1, then move along to the right until the column headed by the third figure (8) is reached and read off the mantissa (93349). Add to this number the mean difference for 3 (the fourth figure of the number) which is 15, and the result is 93364. As the characteristic of 858.3 is 2, log 858.3 = 2.93364. Similarly, log 0.8583 = $\bar{1}$.93364. To find the logarithm of a five figure number, e.g. log 858.38, interpolate linearly between the mean difference for 3 (15) and 4 (20), i.e. (8/10)(20–15) = 4, and, therefore, log 858.38 = 2.93368.

2. **Antilogarithms.** The number (x) corresponding to a given logarithm of x (i.e. its antilogarithm) may be found from Table 2. To find antilog 2.93364, say, ignore the characteristic, and find antilog 0.9336 = 85823. Linear interpolation between the mean difference for 6 (119) and 7 (139) gives (4/10)(139 – 119) = 8, and, therefore, antilog 0.93364 = 85831. Taking account of the characteristic, antilog 2.93364 = 858.31. Similarly, antilog $\bar{3}$.93364 = 0.0085831.

3. **Multiplication.** If x and y are two positive numbers, $\log xy = \log x + \log y$, and, therefore, the product xy is equal to the antilog of the sum of $\log x$ and $\log y$, e.g. log (48.36 x 0.09861) = log 48.36 + log 0.09861 = 1.68449 + $\bar{2}$.99392 = 1.68449 + 0.99392 + $\bar{2}$ = 2.67841 + $\bar{2}$ = 0.67841, and, therefore, antilog 0.67841 = 4.7688, i.e., 48.36 x 0.09861 = 4.7688.

4. **Division.** If x and y are two positive numbers, $\log (x/y) = \log x - \log y$, and, therefore, the quotient x/y is equal to the antilog of ($\log x - \log y$), e.g. log (34.72/0.08156) = log 34.72 – log 0.08156 = 1.54057 – $\bar{2}$.91148 = 1.54057 + 2 – 0.91148 = 3.54057 – 0.91148 = 2.62909 (when subtracting a logarithm the sign of the characteristic and the mantissa is changed), and, therefore, antilog 2.62909 = 425.69, i.e., 34.72/0.08156 = 425.69.

5. **Powers and roots.** $\log x^n = n \log x$, and, therefore, x^n is equal to the antilog of ($n \log x$), e.g. $\log 0.3671^3$ = 3 log 0.3671 = 3($\bar{1}$.56479) = $\bar{3}$ + 1.69437 = $\bar{2}$.69437, and, therefore, antilog $\bar{2}$.69437 = 0.049473, i.e., 0.3671^3 = 0.049473.

Similarly, $\log \sqrt[n]{x} = \log x^{1/n} = 1/n \log x$, and, therefore, $\sqrt[n]{x}$ is equal to the antilog of ($1/n \log x$), e.g. $\log \sqrt[3]{2375}$ = $\frac{1}{3}$log 2375 = $\frac{1}{3}$(3.37568) = 1.12523, and, therefore, antilog 1.12523 = 13.342, i.e., $\sqrt[3]{2375}$ = 13.342. An adjustment is required when the characteristic is negative, e.g. $\log \sqrt[3]{0.02375}$ = $\frac{1}{3}$ log 0.02375 = $\frac{1}{3}$($\bar{2}$.37568) = $\frac{1}{3}$($\bar{3}$ + 1.37568) = $\bar{1}$.45856, and, therefore, antilog $\bar{1}$.45856 = 0.28745, i.e., $\sqrt[3]{0.02375}$ = 0.28745.

6. **Logarithms to base e.** Table 3 gives the characteristic and the mantissa of logarithms to base e of numbers between 1 and 10, e.g. $\log_e$ 7.3462 = 1.99419 (i.e. $\log_e$ 7.346 = 1.99416, and the interpolation for the last figure (2) between the mean difference for 6 (82) and 7 (95) gives (2/10)(95 – 82) = 3).

To find the number (x) whose logarithm to base e is given (i.e. to find its antilogarithm), locate the logarithm in the body of the table and read off the number on the axes, e.g. if $\log_e x$ = 2.23688, the value in the body of the table closest to it is 2.23645 = $\log_e$ 9.36, and the difference between the two numbers is 43 indicating that the third decimal of x is 4, and, therefore, x = 9.364 = antilog 2.23688.

If $x > 10$ use the multiplication rule, e.g. $\log_e 873 = \log_e 8.73 \times 10^2 = \log_e 8.73 + \log_e 10^2$. Log 8.73 can be found from Table 3 directly, and $\log_e 10^2$ = 4.60517 is given at the foot of Table 3. If $x < 1$ proceed similarly, e.g. $\log_e 0.0253 = \log_e 2.53 \times 10^{-2} = \log_e 2.53 + \log_e 10^{-2}$, and $\log_e 10^{-2}$ = $\bar{5}$.39483 can be found at the foot of Table 3.

7. **Switching the base of logarithms.** To convert logarithms to base 10 to logarithms to base e, and vice versa, use the relationships: $\log_e x = \log_e 10 \,.\, \log_{10} x = 2.30258 \log_{10} x$, and $\log_{10} x = \log_{10} e \,.\, \log_e x = 0.43429 \log_e x$.

Table 1. Logarithms (base 10)

x→ ↓	0	1	2	3	4	5	6	7	8	9	Mean differences 1	2	3	4	5	6	7	8	9
10	00000	00432	00860	01284	01703						43	85	128	170	212	255	297	339	382
						02119	02531	02938	03342	03743	41	81	122	162	203	243	283	324	364
11	04139	04532	04922	05308	05690						39	77	116	155	193	232	271	309	348
						06070	06446	06819	07188	07555	37	74	111	148	185	222	259	296	333
12	07918	08279	08636	08991	09342						36	71	107	142	178	213	249	284	319
						09691	10037	10380	10721	11059	34	68	102	137	171	205	239	273	307
13	11394	11727	12057	12385	12710						33	66	99	131	164	197	230	262	295
						13033	13354	13672	13988	14301	32	63	95	127	158	190	221	253	284
14	14613	14922	15229	15534	15836						31	61	92	122	153	183	214	244	274
						16137	16435	16732	17026	17319	30	59	89	118	147	177	206	236	265
15	17609	17898	18184	18469	18752						29	57	86	114	143	171	200	228	256
						19033	19312	19590	19866	20140	28	55	83	111	138	166	193	221	248
16	20412	20683	20952	21219	21484						27	54	80	107	134	161	187	214	241
						21748	22011	22272	22531	22789	26	52	78	104	130	156	182	208	233
17	23045	23300	23553	23805	24055						25	50	76	101	126	151	176	202	227
						24304	24551	24797	25042	25285	25	49	74	98	123	147	171	196	220
18	25527	25768	26007	26245	26482						24	48	72	95	119	143	167	190	214
						26717	26951	27184	27416	27646	23	46	70	93	116	139	162	185	209
19	27875	28103	28330	28556	28780						23	45	68	90	113	136	158	181	203
						29003	29226	29447	29667	29885	22	44	66	88	110	132	154	176	198
20	30103	30320	30535	30750	30963	31175	31387	31597	31806	32015	21	42	64	85	106	127	148	170	191
21	32222	32428	32634	32838	33041	33244	33445	33646	33846	34044	20	40	61	81	101	121	142	162	182
22	34242	34439	34635	34830	35025	35218	35411	35603	35793	35984	19	39	58	77	97	116	135	155	174
23	36173	36361	36549	36736	36922	37107	37291	37475	37658	37840	19	37	56	74	93	111	129	148	166
24	38021	38202	38382	38561	38739	38917	39094	39270	39445	39620	18	36	53	71	89	106	124	142	160
25	39794	39967	40140	40312	40483	40654	40824	40993	41162	41330	17	34	51	68	85	102	119	136	153
26	41497	41664	41830	41996	42160	42325	42488	42651	42813	42975	16	33	49	66	82	98	115	131	148
27	43136	43297	43457	43616	43775	43933	44091	44248	44404	44560	16	32	47	63	79	95	111	126	142
28	44716	44871	45025	45179	45332	45484	45637	45788	45939	46090	15	31	46	61	76	92	107	122	137
29	46240	46389	46538	46687	46835	46982	47129	47276	47422	47567	15	29	44	59	74	88	103	118	133
30	47712	47857	48001	48144	48287	48430	48572	48714	48855	48996	14	29	43	57	71	85	100	114	128
31	49136	49276	49415	49554	49693	49831	49969	50106	50243	50379	14	28	41	55	69	83	97	110	124
32	50515	50651	50786	50920	51055	51188	51322	51455	51587	51720	13	27	40	54	67	80	94	107	120
33	51851	51983	52114	52244	52375	52504	52634	52763	52892	53020	13	26	39	52	65	78	91	104	117
34	53148	53275	53403	53529	53656	53782	53908	54033	54158	54283	13	25	38	50	63	76	88	101	113
35	54407	54531	54654	54777	54900	55023	55145	55267	55388	55509	12	24	37	49	61	73	86	98	110
36	55630	55751	55871	55991	56110	56229	56348	56467	56585	56703	12	24	36	48	60	71	83	95	107
37	56820	56937	57054	57171	57287	57403	57519	57634	57749	57864	12	23	35	46	58	70	81	93	104
38	57978	58092	58206	58320	58433	58546	58659	58771	58883	58995	11	23	34	45	56	68	79	90	102
39	59106	59218	59329	59439	59550	59660	59770	59879	59988	60097	11	22	33	44	55	66	77	88	99
40	60206	60314	60423	60531	60638	60746	60853	60959	61066	61172	11	21	32	43	54	64	75	86	97
41	61278	61384	61490	61595	61700	61805	61909	62014	62118	62221	10	21	31	42	52	63	73	84	94
42	62325	62428	62531	62634	62737	62839	62941	63043	63144	63246	10	20	31	41	51	61	72	82	92
43	63347	63448	63548	63649	63749	63849	63949	64048	64147	64246	10	20	30	40	50	60	70	80	90
44	64345	64444	64542	64640	64738	64836	64933	65031	65128	65225	10	20	29	39	49	59	68	78	88
45	65321	65418	65514	65610	65706	65801	65896	65992	66087	66181	10	19	29	38	48	57	67	76	86
46	66276	66370	66464	66558	66652	66745	66839	66932	67025	67117	9	19	28	37	47	56	65	75	84
47	67210	67302	67394	67486	67578	67669	67761	67852	67943	68034	9	18	27	37	46	55	64	73	82
48	68124	68215	68305	68395	68485	68574	68664	68753	68842	68931	9	18	27	36	45	54	63	72	81
49	69020	69108	69197	69285	69373	69461	69548	69636	69723	69810	9	18	26	35	44	53	61	70	79
50	69897	69984	70070	70157	70243	70329	70415	70501	70586	70672	9	17	26	34	43	52	60	69	77
51	70757	70842	70927	71012	71096	71181	71265	71349	71433	71517	8	17	25	34	42	51	59	67	76
52	71600	71684	71767	71850	71933	72016	72099	72181	72263	72346	8	17	25	33	41	50	58	66	74
53	72428	72509	72591	72673	72754	72835	72916	72997	73078	73159	8	16	24	32	41	49	57	65	73
54	73239	73320	73400	73480	73560	73640	73719	73799	73878	73957	8	16	24	32	40	48	56	64	72
55	74036	74115	74194	74273	74351	74429	74507	74586	74663	74741	8	16	23	31	39	47	55	63	70
56	74819	74896	74974	75051	75128	75205	75282	75358	75435	75511	8	15	23	31	38	46	54	62	69
57	75587	75664	75740	75815	75891	75967	76042	76118	76193	76268	8	15	23	30	38	45	53	60	68
58	76343	76418	76492	76567	76641	76716	76790	76864	76938	77012	7	15	22	30	37	45	52	59	67
59	77085	77159	77232	77305	77379	77452	77525	77597	77670	77743	7	15	22	29	37	44	51	58	66

Table 1 – *continued*

x→ ↓	0	1	2	3	4	5	6	7	8	9	Mean differences								
											1	2	3	4	5	6	7	8	9
60	77815	77887	77960	78032	78104	78176	78247	78319	78390	78462	7	14	22	29	36	43	50	57	65
61	78533	78604	78675	78746	78817	78888	78958	79029	79099	79169	7	14	21	28	35	42	49	57	64
62	79239	79309	79379	79449	79518	79588	79657	79727	79796	79865	7	14	21	28	35	42	49	56	63
63	79934	80003	80072	80140	80209	80277	80346	80414	80482	80550	7	14	21	27	34	41	48	55	62
64	80618	80686	80754	80821	80889	80956	81023	81090	81158	81224	7	13	20	27	34	40	47	54	61
65	81291	81358	81425	81491	81558	81624	81690	81757	81823	81889	7	13	20	27	33	40	46	53	60
66	81954	82020	82086	82151	82217	82282	82347	82413	82478	82543	7	13	20	26	33	39	46	52	59
67	82607	82672	82737	82802	82866	82930	82995	83059	83123	83187	6	13	19	26	32	39	45	51	58
68	83251	83315	83378	83442	83506	83569	83632	83696	83759	83822	6	13	19	25	32	38	44	51	57
69	83885	83948	84011	84073	84136	84198	84261	84323	84386	84448	6	13	19	25	31	38	44	50	56
70	84510	84572	84634	84696	84757	84819	84880	84942	85003	85065	6	12	18	25	31	37	43	49	55
71	85126	85187	85248	85309	85370	85431	85491	85552	85612	85673	6	12	18	24	30	36	43	49	55
72	85733	85794	85854	85914	85974	86034	86094	86153	86213	86273	6	12	18	24	30	36	42	48	54
73	86332	86392	86451	86510	86570	86629	86688	86747	86806	86864	6	12	18	24	30	35	41	47	53
74	86923	86982	87040	87099	87157	87216	87274	87332	87390	87448	6	12	17	23	29	35	41	47	52
75	87506	87564	87622	87679	87737	87795	87852	87910	87967	88024	6	12	17	23	29	35	40	46	52
76	88081	88138	88195	88252	88309	88366	88423	88480	88536	88593	6	11	17	23	28	34	40	45	51
77	88649	88705	88762	88818	88874	88930	88986	89042	89098	89154	6	11	17	22	28	34	39	45	50
78	89209	89265	89321	89376	89432	89487	89542	89597	89653	89708	6	11	17	22	28	33	39	44	50
79	89763	89818	89873	89927	89982	90037	90091	90146	90200	90255	5	11	16	22	27	33	38	44	49
80	90309	90363	90417	90472	90526	90580	90634	90687	90741	90795	5	11	16	22	27	32	38	43	49
81	90849	90902	90956	91009	91062	91116	91169	91222	91275	91328	5	11	16	21	27	32	37	43	48
82	91381	91434	91487	91540	91593	91645	91698	91751	91803	91855	5	11	16	21	26	32	37	42	47
83	91908	91960	92012	92065	92117	92169	92221	92273	92324	92376	5	10	16	21	26	31	36	42	47
84	92428	92480	92531	92583	92634	92686	92737	92788	92840	92891	5	10	15	21	26	31	36	41	46
85	92942	92993	93044	93095	93146	93197	93247	93298	93349	93399	5	10	15	20	25	30	36	41	46
86	93450	93500	93551	93601	93651	93702	93752	93802	93852	93902	5	10	15	20	25	30	35	40	45
87	93952	94002	94052	94101	94151	94201	94250	94300	94349	94399	5	10	15	20	25	30	35	40	45
88	94448	94498	94547	94596	94645	94694	94743	94792	94841	94890	5	10	15	20	25	29	34	39	44
89	94939	94988	95036	95085	95134	95182	95231	95279	95328	95376	5	10	15	19	24	29	34	39	44
90	95424	95472	95521	95569	95617	95665	95713	95761	95809	95856	5	10	14	19	24	29	34	38	43
91	95904	95952	95999	96047	96095	96142	96190	96237	96284	96332	5	9	14	19	24	28	33	38	43
92	96379	96426	96473	96520	96567	96614	96661	96708	96755	96802	5	9	14	19	23	28	33	38	42
93	96848	96895	96942	96988	97035	97081	97128	97174	97220	97267	5	9	14	19	23	28	33	37	42
94	97313	97359	97405	97451	97497	97543	97589	97635	97681	97727	5	9	14	18	23	28	32	37	41
95	97772	97818	97864	97909	97955	98000	98046	98091	98137	98182	5	9	14	18	23	27	32	36	41
96	98227	98272	98318	98363	98408	98453	98498	98543	98588	98632	5	9	14	18	23	27	32	36	41
97	98677	98722	98767	98811	98856	98900	98945	98989	99034	99078	4	9	13	18	22	27	31	36	40
98	99123	99167	99211	99255	99300	99344	99388	99432	99476	99520	4	9	13	18	22	26	31	35	40
99	99564	99607	99651	99695	99739	99782	99826	99870	99913	99957	4	9	13	17	22	26	31	35	39

Table 2. Antilogarithms (base 10)

log x ↓ →	0	1	2	3	4	5	6	7	8	9	Mean differences								
											1	2	3	4	5	6	7	8	9
.00	10000	10023	10046	10069	10093	10116	10139	10162	10186	10209	2	5	7	9	12	14	16	19	21
.01	10233	10257	10280	10304	10328	10351	10375	10399	10423	10447	2	5	7	10	12	14	17	19	21
.02	10471	10495	10520	10544	10568	10593	10617	10641	10666	10691	2	5	7	10	12	15	17	20	22
.03	10715	10740	10765	10789	10814	10839	10864	10889	10914	10940	2	5	7	10	12	15	17	20	22
.04	10965	10990	11015	11041	11066	11092	11117	11143	11169	11194	3	5	8	10	13	15	18	20	23
.05	11220	11246	11272	11298	11324	11350	11376	11402	11429	11455	3	5	8	10	13	16	18	21	24
.06	11482	11508	11535	11561	11588	11614	11641	11668	11695	11722	3	5	8	11	13	16	19	21	24
.07	11749	11776	11803	11830	11858	11885	11912	11940	11967	11995	3	5	8	11	14	16	19	22	25
.08	12023	12050	12078	12106	12134	12162	12190	12218	12246	12274	3	6	8	11	14	17	20	22	25
.09	12303	12331	12359	12388	12417	12445	12474	12503	12531	12560	3	6	9	11	14	17	20	23	26
.10	12589	12618	12647	12677	12706	12735	12764	12794	12823	12853	3	6	9	12	15	18	21	23	26
.11	12882	12912	12942	12972	13002	13032	13062	13092	13122	13152	3	6	9	12	15	18	21	24	27
.12	13183	13213	13243	13274	13305	13335	13366	13397	13428	13459	3	6	9	12	15	18	21	25	28
.13	13490	13521	13552	13583	13614	13646	13677	13709	13740	13772	3	6	9	13	16	19	22	25	28
.14	13804	13836	13868	13900	13932	13964	13996	14028	14060	14093	3	6	10	13	16	19	22	26	29
.15	14125	14158	14191	14223	14256	14289	14322	14355	14388	14421	3	7	10	13	16	20	23	26	30
.16	14454	14488	14521	14555	14588	14622	14655	14689	14723	14757	3	7	10	13	17	20	24	27	30
.17	14791	14825	14859	14894	14928	14962	14997	15031	15066	15101	3	7	10	14	17	21	24	28	31
.18	15136	15171	15205	15241	15276	15311	15346	15382	15417	15453	4	7	11	14	18	21	25	28	32
.19	15488	15524	15560	15596	15631	15668	15704	15740	15776	15812	4	7	11	14	18	22	25	29	32
.20	15849	15885	15922	15959	15996	16032	16069	16106	16144	16181	4	7	11	15	18	22	26	30	33
.21	16218	16255	16293	16331	16368	16406	16444	16482	16520	16558	4	8	11	15	19	23	26	30	34
.22	16596	16634	16672	16711	16749	16788	16827	16866	16904	16943	4	8	12	15	19	23	27	31	35
.23	16982	17022	17061	17100	17140	17179	17219	17258	17298	17338	4	8	12	16	20	24	28	32	36
.24	17378	17418	17458	17498	17539	17579	17620	17660	17701	17742	4	8	12	16	20	24	28	32	36
.25	17783	17824	17865	17906	17947	17989	18030	18072	18113	18155	4	8	12	17	21	25	29	33	37
.26	18197	18239	18281	18323	18365	18408	18450	18493	18535	18578	4	8	13	17	21	25	30	34	38
.27	18621	18664	18707	18750	18793	18836	18880	18923	18967	19011	4	9	13	17	22	26	30	35	39
.28	19055	19099	19143	19187	19231	19275	19320	19364	19409	19454	4	9	13	18	22	27	31	35	40
.29	19498	19543	19588	19634	19679	19724	19770	19815	19861	19907	5	9	14	18	23	27	32	36	41
.30	19953	19999	20045	20091	20137	20184	20230	20277	20324	20370	5	9	14	19	23	28	33	37	42
.31	20417	20464	20512	20559	20606	20654	20701	20749	20797	20845	5	10	14	19	24	29	33	38	43
.32	20893	20941	20989	21038	21086	21135	21184	21232	21281	21330	5	10	15	19	24	29	34	39	44
.33	21380	21429	21478	21528	21577	21627	21677	21727	21777	21827	5	10	15	20	25	30	35	40	45
.34	21878	21928	21979	22029	22080	22131	22182	22233	22284	22336	5	10	15	20	25	31	36	41	46
.35	22387	22439	22491	22542	22594	22646	22699	22751	22803	22856	5	10	16	21	26	31	36	42	47
.36	22909	22961	23014	23067	23121	23174	23227	23281	23335	23388	5	11	16	21	27	32	37	43	48
.37	23442	23496	23550	23605	23659	23714	23768	23823	23878	23933	5	11	16	22	27	33	38	44	49
.38	23988	24044	24099	24155	24210	24266	24322	24378	24434	24491	6	11	17	22	28	34	39	45	50
.39	24547	24604	24660	24717	24774	24831	24889	24946	25003	25061	6	11	17	23	29	34	40	46	51
.40	25119	25177	25235	25293	25351	25410	25468	25527	25586	25645	6	12	18	23	29	35	41	47	53
.41	25704	25763	25823	25882	25942	26002	26062	26122	26182	26242	6	12	18	24	30	36	42	48	54
.42	26303	26363	26424	26485	26546	26607	26669	26730	26792	26853	6	12	18	24	31	37	43	49	55
.43	26915	26977	27040	27102	27164	27227	27290	27353	27416	27479	6	13	19	25	31	38	44	50	56
.44	27542	27606	27669	27733	27797	27861	27925	27990	28054	28119	6	13	19	26	32	38	45	51	58
.45	28184	28249	28314	28379	28445	28510	28576	28642	28708	28774	7	13	20	26	33	39	46	53	59
.46	28840	28907	28973	29040	29107	29174	29242	29309	29376	29444	7	13	20	27	34	40	47	54	60
.47	29512	29580	29648	29717	29785	29854	29923	29992	30061	30130	7	14	21	27	34	41	48	55	62
.48	30200	30269	30339	30409	30479	30549	30620	30690	30761	30832	7	14	21	28	35	42	49	56	63
.49	30903	30974	31046	31117	31189	31261	31333	31405	31477	31550	7	14	22	29	36	43	50	58	65
.50	31623	31696	31769	31842	31915	31989	32063	32137	32211	32285	7	15	22	29	37	44	52	59	66
.51	32359	32434	32509	32584	32659	32734	32810	32885	32961	33037	8	15	23	30	38	45	53	60	68
.52	33113	33189	33266	33343	33420	33497	33574	33651	33729	33806	8	15	23	31	39	46	54	62	69
.53	33884	33963	34041	34119	34198	34277	34356	34435	34514	34594	8	16	24	32	39	47	55	63	71
.54	34674	34754	34834	34914	34995	35075	35156	35237	35318	35400	8	16	24	32	40	48	57	65	73
.55	35481	35563	35645	35727	35810	35892	35975	36058	36141	36224	8	17	25	33	41	50	58	66	74
.56	36308	36392	36475	36559	36644	36728	36813	36898	36983	37068	8	17	25	34	42	51	59	68	76
.57	37154	37239	37325	37411	37497	37584	37670	37757	37844	37931	9	17	26	35	43	52	61	69	78
.58	38019	38107	38194	38282	38371	38459	38548	38637	38726	38815	9	18	27	35	44	53	62	71	80
.59	38905	38994	39084	39174	39264	39355	39446	39537	39628	39719	9	18	27	36	45	54	63	72	82

Table 2 – *continued*

log x ↓→	0	1	2	3	4	5	6	7	8	9	Mean differences								
											1	2	3	4	5	6	7	8	9
.60	39811	39902	39994	40087	40179	40272	40365	40458	40551	40644	9	19	28	37	46	56	65	74	83
.61	40738	40832	40926	41020	41115	41210	41305	41400	41495	41591	9	19	28	38	47	57	66	76	85
.62	41687	41783	41879	41976	42073	42170	42267	42364	42462	42560	10	19	29	39	49	58	68	78	87
.63	42658	42756	42855	42954	43053	43152	43251	43351	43451	43551	10	20	30	40	50	60	70	79	89
.64	43652	43752	43853	43954	44055	44157	44259	44361	44463	44566	10	20	30	41	51	61	71	81	91
.65	44668	44771	44875	44978	45082	45186	45290	45394	45499	45604	10	21	31	42	52	62	73	83	94
.66	45709	45814	45920	46026	46132	46238	46345	46452	46559	46666	11	21	32	43	53	64	75	85	96
.67	46774	46881	46989	47098	47206	47315	47424	47534	47643	47753	11	22	33	44	54	65	76	87	98
.68	47863	47973	48084	48195	48306	48417	48529	48641	48753	48865	11	22	33	45	56	67	78	89	100
.69	48978	49091	49204	49317	49431	49545	49659	49774	49888	50003	11	23	34	46	57	68	80	91	103
.70	50119	50234	50350	50466	50582	50699	50816	50933	51050	51168	12	23	35	47	58	70	82	93	105
.71	51286	51404	51523	51642	51761	51880	52000	52119	52240	52360	12	24	36	48	60	72	84	96	108
.72	52481	52602	52723	52845	52966	53088	53211	53333	53456	53580	12	24	37	49	61	73	86	98	110
.73	53703	53827	53951	54075	54200	54325	54450	54576	54702	54828	12	25	37	50	63	75	88	100	113
.74	54954	55081	55208	55335	55463	55590	55719	55847	55976	56105	13	26	38	51	64	77	90	102	115
.75	56234	56364	56494	56624	56754	56885	57016	57148	57280	57412	13	26	39	52	65	79	92	105	118
.76	57544	57677	57810	57943	58076	58210	58345	58479	58614	58749	13	27	40	54	67	80	94	107	121
.77	58884	59020	59156	59293	59429	59566	59704	59841	59979	60117	14	27	41	55	69	82	96	110	123
.78	60256	60395	60534	60674	60814	60954	61094	61235	61376	61518	14	28	42	56	70	84	98	112	126
.79	61660	61802	61944	62087	62230	62373	62517	62661	62806	62951	14	29	43	57	72	86	101	115	129
.80	63096	63241	63387	63433	63680	63826	63973	64121	64269	64417	15	29	44	59	73	88	103	118	132
.81	64565	64714	64863	65013	65163	65313	65464	65615	65766	65917	15	30	45	60	75	90	105	120	135
.82	66069	66222	66374	66527	66681	66834	66988	67143	67298	67453	15	31	46	62	77	92	108	123	138
.83	67608	67764	67920	68077	68234	68391	68549	68707	68865	69024	16	31	47	63	79	94	110	126	142
.84	69183	69343	69502	69663	69823	69984	70146	70307	70469	70632	16	32	48	64	81	97	113	129	145
.85	70795	70958	71121	71285	71450	71614	71779	71945	72111	72277	16	33	49	66	82	99	115	132	148
.86	72444	72611	72778	72946	73114	73282	73451	73621	73790	73961	17	34	51	67	84	101	118	135	152
.87	74131	74302	74473	74645	74817	74989	75162	75336	75509	75683	17	35	52	69	86	104	121	138	155
.88	75858	76033	76208	76384	76560	76736	76913	77090	77268	77446	18	35	53	71	88	106	124	141	159
.89	77625	77804	77983	78163	78343	78524	78705	78886	79068	79250	18	36	54	72	90	108	127	145	163
.90	79433	79616	79799	79983	80168	80353	80538	80724	80910	81096	18	37	55	74	92	111	129	148	167
.91	81283	81470	81658	81846	82035	82224	82414	82604	82794	82985	19	38	57	76	95	114	132	151	170
.92	83176	83368	83560	83753	83946	84140	84333	84528	84723	84918	19	39	58	77	97	116	136	155	174
.93	85114	85310	85507	85704	85901	86099	86298	86497	86696	86896	20	40	59	79	99	119	139	159	178
.94	87096	87297	87498	87700	87902	88105	88308	88512	88716	88920	20	41	61	81	101	122	142	162	183
.95	89125	89331	89536	89743	89950	90157	90365	90573	90782	90991	21	41	62	83	104	125	145	166	187
.96	91201	91411	91622	91833	92045	92257	92470	92683	92897	93111	21	42	64	85	106	127	149	170	191
.97	93325	93541	93756	93972	94189	94406	94624	94842	95060	95280	22	43	65	87	109	130	152	174	196
.98	95499	95719	95940	96161	96383	96605	96828	97051	97275	97499	22	44	67	89	111	133	156	178	200
.99	97724	97949	98175	98401	98628	98855	99083	99312	99541	99770	23	45	68	91	114	137	159	182	205

Table 3. Logarithms (base *e*)

x→ ↓	0	1	2	3	4	5	6	7	8	9	Mean differences								
											1	2	3	4	5	6	7	8	9
1.0	0.00000	0.00995	0.01980	0.02956	0.03922	0.04879	0.05827	0.06766	0.07696	0.08618	96	191	287	382	478	573	668	763	858
1.1	0.09531	0.10436	0.11333	0.12222	0.13103	0.13976	0.14842	0.15700	0.16551	0.17395	87	175	262	349	436	523	610	697	783
1.2	0.18232	0.19062	0.19885	0.20701	0.21511	0.22314	0.23111	0.23902	0.24686	0.25464	80	161	241	321	401	481	561	641	721
1.3	0.26236	0.27003	0.27763	0.28518	0.29267	0.30010	0.30748	0.31481	0.32208	0.32930	74	149	223	297	371	445	519	593	667
1.4	0.33647	0.34359	0.35066	0.35767	0.36464	0.37156	0.37844	0.38526	0.39204	0.39878	69	138	207	277	346	415	483	552	621
1.5	0.40547	0.41211	0.41871	0.42527	0.43178	0.43825	0.44469	0.45108	0.45742	0.46373	65	129	194	259	323	388	452	517	581
1.6	0.47000	0.47623	0.48243	0.48858	0.49470	0.50078	0.50682	0.51282	0.51879	0.52473	61	122	182	243	304	364	425	485	546
1.7	0.53063	0.53649	0.54232	0.54812	0.55389	0.55962	0.56531	0.57098	0.57661	0.58222	57	115	172	229	286	343	400	458	515
1.8	0.58779	0.59333	0.59884	0.60432	0.60977	0.61519	0.62058	0.62594	0.63127	0.63658	54	108	163	217	271	325	379	433	487
1.9	0.64185	0.64710	0.65233	0.65752	0.66269	0.66783	0.67294	0.67803	0.68310	0.68813	51	103	154	205	257	308	359	411	462
2.0	0.69315	0.69813	0.70310	0.70804	0.71295	0.71784	0.72271	0.72755	0.73237	0.73716	49	98	147	195	244	293	342	391	439
2.1	0.74194	0.74669	0.75142	0.75612	0.76081	0.76547	0.77011	0.77473	0.77932	0.78390	47	93	140	186	233	279	326	372	419
2.2	0.78846	0.79299	0.79751	0.80200	0.80648	0.81093	0.81536	0.81978	0.82418	0.82855	45	89	134	178	223	267	311	356	400
2.3	0.83291	0.83725	0.84157	0.84587	0.85015	0.85442	0.85866	0.86289	0.86710	0.87129	43	85	128	170	213	256	298	341	383
2.4	0.87547	0.87963	0.88377	0.88789	0.89200	0.89609	0.90016	0.90422	0.90826	0.91228	41	82	123	163	204	245	286	327	367
2.5	0.91629	0.92028	0.92426	0.92822	0.93216	0.93609	0.94001	0.94391	0.94779	0.95166	39	79	118	157	196	236	275	314	353
2.6	0.95551	0.95935	0.96317	0.96698	0.97078	0.97456	0.97833	0.98208	0.98582	0.98954	38	76	113	151	189	227	264	302	340
2.7	0.99325	0.99695	1.00063	1.00430	1.00796	1.01160	1.01523	1.01885	1.02245	1.02604	36	73	109	146	182	218	255	291	327
2.8	1.02962	1.03318	1.03674	1.04028	1.04380	1.04732	1.05082	1.05431	1.05779	1.06126	35	70	105	141	176	211	246	281	316
2.9	1.06471	1.06815	1.07158	1.07500	1.07841	1.08181	1.08519	1.08856	1.09192	1.09527	34	68	102	136	170	204	237	271	305
3.0	1.09861	1.10194	1.10526	1.10856	1.11186	1.11514	1.11841	1.12168	1.12493	1.12817	33	66	98	131	164	197	230	262	295
3.1	1.13140	1.13462	1.13783	1.14103	1.14422	1.14740	1.15057	1.15373	1.15688	1.16002	32	64	95	127	159	191	222	254	286
3.2	1.16315	1.16627	1.16938	1.17248	1.17557	1.17865	1.18173	1.18479	1.18784	1.19089	31	62	92	123	154	185	216	246	277
3.3	1.19392	1.19695	1.19996	1.20297	1.20597	1.20896	1.21194	1.21491	1.21788	1.22083	30	60	90	120	149	179	209	239	269
3.4	1.22378	1.22671	1.22964	1.23256	1.23547	1.23837	1.24127	1.24415	1.24703	1.25990	29	58	87	116	145	174	203	232	261
3.5	1.25276	1.25562	1.25846	1.26130	1.26413	1.26695	1.26976	1.27257	1.27536	1.27815	28	56	85	113	141	169	197	225	254
3.6	1.28093	1.28371	1.28647	1.28923	1.29198	1.29473	1.29746	1.30019	1.30291	1.30563	27	55	82	110	137	164	192	219	247
3.7	1.30833	1.31103	1.31372	1.31641	1.31909	1.32176	1.32442	1.32708	1.32972	1.33237	27	53	80	107	133	160	187	213	240
3.8	1.33500	1.33763	1.34025	1.34286	1.34547	1.34807	1.35067	1.35325	1.35584	1.35841	26	52	78	104	130	156	182	208	234
3.9	1.36098	1.36354	1.36609	1.36864	1.37118	1.37372	1.37624	1.37877	1.38128	1.38379	25	51	76	101	127	152	177	203	228
4.0	1.38629	1.38879	1.39128	1.39377	1.39624	1.39872	1.40118	1.40364	1.40610	1.40854	25	49	74	99	124	148	173	198	222
4.1	1.41099	1.41342	1.41585	1.41828	1.42070	1.42311	1.42552	1.42792	1.43031	1.43270	24	48	72	96	121	145	169	193	217
4.2	1.43508	1.43746	1.43984	1.44220	1.44456	1.44692	1.44927	1.45161	1.45395	1.45629	24	47	71	94	118	141	165	188	212
4.3	1.45862	1.46094	1.46326	1.46557	1.46787	1.47018	1.47247	1.47476	1.47705	1.47933	23	46	69	92	115	138	161	184	207
4.4	1.48160	1.48387	1.48614	1.48840	1.49065	1.49290	1.49515	1.49739	1.49962	1.50185	22	45	67	90	112	135	157	180	202
4.5	1.50408	1.50630	1.50851	1.51072	1.51293	1.51513	1.51732	1.51951	1.52170	1.52388	22	44	66	88	110	132	154	176	198
4.6	1.52606	1.52823	1.53039	1.53256	1.53471	1.53687	1.53902	1.54116	1.54330	1.54543	22	43	65	86	108	129	151	172	194
4.7	1.54756	1.54969	1.55181	1.55393	1.55604	1.55814	1.56025	1.56235	1.56444	1.56653	21	42	63	84	105	126	147	168	190
4.8	1.56862	1.57070	1.57277	1.57485	1.57691	1.57898	1.58104	1.58309	1.58515	1.58719	21	41	62	83	103	124	144	165	186
4.9	1.58924	1.59127	1.59331	1.59534	1.59737	1.59939	1.60141	1.60342	1.60543	1.60744	20	40	61	81	101	121	141	162	182
5.0	1.60944	1.61144	1.61343	1.61542	1.61741	1.61939	1.62137	1.62334	1.62531	1.62728	20	40	59	79	99	119	139	158	178
5.1	1.62924	1.63120	1.63315	1.63511	1.63705	1.63900	1.64094	1.64287	1.64481	1.64673	19	39	58	78	97	117	136	155	175
5.2	1.64866	1.65058	1.65250	1.65441	1.65632	1.65823	1.66013	1.66203	1.66393	1.66582	19	38	57	76	95	114	133	152	171
5.3	1.66771	1.66959	1.67147	1.67335	1.67523	1.67710	1.67896	1.68083	1.68269	1.68455	19	37	56	75	94	112	131	150	168
5.4	1.68640	1.68825	1.69010	1.69194	1.69378	1.69562	1.69745	1.69928	1.70111	1.70293	18	37	55	73	92	110	128	147	165
5.5	1.70475	1.70656	1.70838	1.71019	1.71199	1.71380	1.71560	1.71740	1.71919	1.72098	18	36	54	72	90	108	126	144	162
5.6	1.72277	1.72455	1.72633	1.72811	1.72988	1.73166	1.73342	1.73519	1.73695	1.73871	18	35	53	71	89	106	124	142	159
5.7	1.74047	1.74222	1.74397	1.74572	1.74746	1.74920	1.75094	1.75267	1.75440	1.75613	17	35	52	70	87	104	122	139	157
5.8	1.75786	1.75958	1.76130	1.76302	1.76473	1.76644	1.76815	1.76985	1.77156	1.77326	17	34	51	68	86	103	120	137	154
5.9	1.77495	1.77665	1.77834	1.78002	1.78171	1.78339	1.78507	1.78675	1.78842	1.79009	17	34	50	67	84	101	118	134	151
6.0	1.79176	1.79342	1.79509	1.79675	1.79840	1.80006	1.80171	1.80336	1.80500	1.80665	17	33	50	66	83	99	116	132	149
6.1	1.80829	1.80993	1.81156	1.81319	1.81482	1.81645	1.81808	1.81970	1.82132	1.82294	16	33	49	65	81	98	114	130	146
6.2	1.82455	1.82616	1.82777	1.82938	1.83098	1.83258	1.83418	1.83578	1.83737	1.83896	16	32	48	64	80	96	112	128	144
6.3	1.84055	1.84214	1.84372	1.84530	1.84688	1.84845	1.85003	1.85160	1.85317	1.85473	16	32	47	63	79	95	110	126	142
6.4	1.85630	1.85786	1.85942	1.86097	1.86253	1.86408	1.86563	1.86718	1.86872	1.87026	16	31	47	62	78	93	109	124	140
6.5	1.87180	1.87334	1.87487	1.87641	1.87794	1.87947	1.88099	1.88251	1.88403	1.88555	15	31	46	61	76	92	107	122	137
6.6	1.88707	1.88858	1.89010	1.89160	1.89311	1.89462	1.89612	1.89762	1.89912	1.90061	15	30	45	60	75	90	105	120	135
6.7	1.90211	1.90360	1.90509	1.90658	1.90806	1.90954	1.91102	1.91250	1.91398	1.91545	15	30	44	59	74	89	104	119	133
6.8	1.91692	1.91839	1.91986	1.92132	1.92279	1.92425	1.92571	1.92716	1.92862	1.93007	15	29	44	58	73	88	102	117	131
6.9	1.93152	1.93297	1.93442	1.93586	1.93730	1.93874	1.94018	1.94162	1.94305	1.94448	14	29	43	58	72	86	101	115	130

n	1	2	3	4	5	6	7	8	9
$\log_e 10^n$	2.30259	4.60517	6.90776	9.21034	11.51293	13.81551	16.11810	18.42068	20.72327

Table 3 – *continued*

$x\rightarrow$ ↓	0	1	2	3	4	5	6	7	8	9	Mean differences 1	2	3	4	5	6	7	8	9
7.0	1.94591	1.94734	1.94876	1.95019	1.95161	1.95303	1.95445	1.95586	1.95727	1.95869	14	28	43	57	71	85	99	113	128
7.1	1.96009	1.96150	1.96291	1.96431	1.96571	1.96711	1.96851	1.96991	1.97130	1.97269	14	28	42	56	70	84	98	112	126
7.2	1.97408	1.97547	1.97685	1.97824	1.97962	1.98100	1.98238	1.98376	1.98513	1.98650	14	28	41	55	69	83	97	110	124
7.3	1.98787	1.98924	1.99061	1.99198	1.99334	1.99470	1.99606	1.99742	1.99877	2.00013	14	27	41	54	68	82	95	109	122
7.4	2.00148	2.00283	2.00418	2.00353	2.00687	2.00821	2.00956	2.01089	2.01223	2.01357	13	27	40	54	67	81	94	107	121
7.5	2.01490	2.01624	2.01757	2.01890	2.02022	2.02155	2.02287	2.02419	2.02551	2.02683	13	27	40	53	66	79	93	106	119
7.6	2.02815	2.02946	2.03078	2.03209	2.03340	2.03471	2.03601	2.03732	2.03862	2.03992	13	26	39	52	65	78	92	105	118
7.7	2.04122	2.04252	2.04381	2.04511	2.04640	2.04769	2.04898	2.05027	2.05156	2.05284	13	26	39	52	65	77	90	103	116
7.8	2.05412	2.05540	2.05668	2.05796	2.05924	2.06051	2.06179	2.06306	2.06433	2.06560	13	25	38	51	64	76	89	102	115
7.9	2.06686	2.06813	2.06939	2.07065	2.07191	2.07317	2.07443	2.07568	2.07694	2.07819	13	25	38	50	63	75	88	101	113
8.0	2.07944	2.08069	2.08194	2.08318	2.08443	2.08567	2.08691	2.08815	2.08939	2.09063	12	25	37	50	62	75	87	99	112
8.1	2.09186	2.09310	2.09433	2.09556	2.09679	2.09802	2.09924	2.10047	2.10169	2.10291	12	25	37	49	61	74	86	98	110
8.2	2.10413	2.10535	2.10657	2.10779	2.10900	2.11021	2.11142	2.11263	2.11384	2.11505	12	24	36	49	61	73	85	97	109
8.3	2.11626	2.11746	2.11866	2.11986	2.12106	2.12226	2.12346	2.12465	2.12585	2.12704	12	24	36	48	60	72	84	96	108
8.4	2.12823	2.12942	2.13061	2.13180	2.13298	2.13417	2.13535	2.13653	2.13771	2.13889	12	24	36	47	59	71	83	95	107
8.5	2.14007	2.14124	2.14242	2.14359	2.14476	2.14593	2.14710	2.14827	2.14943	2.15060	12	23	35	47	58	70	82	94	105
8.6	1.15176	2.15292	2.15409	2.15524	2.15640	2.15756	2.15871	2.15987	2.16102	2.16217	12	23	35	46	58	69	81	92	104
8.7	2.16332	2.16447	2.16562	2.16677	2.16791	2.16905	2.17020	2.17134	2.17248	2.17361	11	23	34	46	57	69	80	91	103
8.8	2.17475	2.17589	2.17702	2.17816	2.17929	2.18042	2.18155	2.18267	2.18380	2.18493	11	23	34	45	57	68	79	90	102
8.9	2.18605	2.18717	2.18830	2.18942	2.19054	2.19165	2.19277	2.19389	2.19500	2.19611	11	22	34	45	56	67	78	89	101
9.0	2.19722	2.19834	2.19944	2.20055	2.20166	2.20276	2.20387	2.20497	2.20607	2.20717	11	22	33	44	55	66	77	88	99
9.1	2.20827	2.20937	2.21047	2.21157	2.21266	2.21375	2.21485	2.21594	2.21703	2.21812	11	22	33	44	55	66	77	87	98
9.2	2.21920	2.22029	2.22138	2.22246	2.22354	2.22462	2.22570	2.22678	2.22786	2.22894	11	22	32	43	54	65	76	86	97
9.3	2.23001	2.23109	2.23216	2.23324	2.23431	2.23538	2.23645	2.23751	2.23858	2.23965	11	21	32	43	53	64	75	86	96
9.4	2.24071	2.24177	2.24284	2.24390	2.24496	2.24601	2.24707	2.24813	2.24918	2.25024	11	21	32	42	53	64	74	85	95
9.5	2.25129	2.25234	2.25339	2.25444	2.25549	2.25654	2.25759	2.25863	2.25968	2.26072	10	21	31	42	52	63	73	84	94
9.6	2.26176	2.26280	2.26384	2.26488	2.26592	2.26696	2.26799	2.26903	2.27006	2.27109	10	21	31	41	52	62	73	83	93
9.7	2.27213	2.27316	2.27419	2.27521	2.27624	2.27727	2.27829	2.27932	2.28034	2.28136	10	21	31	41	51	62	72	82	92
9.8	2.28238	2.28340	2.28442	2.28544	2.28646	2.28747	2.28849	2.28950	2.29051	2.29152	10	20	30	41	51	61	71	81	91
9.9	2.29253	2.29354	2.29455	2.29556	2.29657	2.29757	2.29858	2.29958	2.30058	2.30158	10	20	30	40	50	60	70	80	90
10.0	2.30259																		

n	1	2	3	4	5	6	7	8	9
$\log_e 10^{-n}$	$\bar{3}.69741$	$\bar{5}.39483$	$\bar{7}.09224$	$\overline{10}.78966$	$\overline{12}.48707$	$\overline{14}.18449$	$\overline{17}.88190$	$\overline{19}.57932$	$\overline{21}.27673$

Table 4. Squares, cubes, square roots and reciprocals

n	n^2	n^3	$\sqrt{n}$	$\sqrt[3]{n}$	$\sqrt{10n}$	$\sqrt[3]{10n}$	$\sqrt[3]{100n}$	$1/n$
1	1	1	1.00000	1.00000	3.16228	2.15442	4.64152	1.00000
2	4	8	1.41421	1.25992	4.47214	2.71439	5.84793	0.50000
3	9	27	1.73205	1.44224	5.47723	3.10720	5.69420	0.33333
4	16	64	2.00000	1.58739	6.32456	3.41991	7.36792	0.25000
5	25	125	2.23607	1.70997	7.07107	3.68398	7.93684	0.20000
6	36	216	2.44949	1.81711	7.74597	3.91481	8.43415	0.16667
7	49	343	2.64575	1.91292	8.36660	4.12123	8.87885	0.14286
8	64	512	2.82843	2.00000	8.94427	4.30881	9.28297	0.12500
9	81	729	3.00000	2.08007	9.48683	4.48134	9.65467	0.11111
10	100	1000	3.16228	2.15442	10.00000	4.64152	9.99977	0.10000
11	121	1331	3.31662	2.22396	10.48809	4.79134	10.32256	0.09091
12	144	1728	3.46410	2.28941	10.95445	4.93235	10.62633	0.08333
13	169	2197	3.60555	2.35131	11.40175	5.06571	10.91367	0.07692
14	196	2744	3.74166	2.41012	11.83216	5.19241	11.18662	0.07143
15	225	3375	3.87298	2.46619	12.24745	5.31320	11.44686	0.06667
16	256	4096	4.00000	2.51982	12.64911	5.42874	11.69578	0.06250
17	289	4913	4.12311	2.57126	13.03840	5.53956	11.93454	0.05882
18	324	5832	4.24264	2.62072	13.41641	5.64612	12.16410	0.05556
19	361	6859	4.35890	2.66838	13.78405	5.74880	12.38531	0.05263
20	400	8000	4.47214	2.71439	14.14214	5.84793	12.59889	0.05000
21	441	9261	4.58258	2.75890	14.49138	5.94382	12.80547	0.04762
22	484	10648	4.69042	2.80201	14.83240	6.03670	13.00558	0.04545
23	529	12167	4.79583	2.84384	15.16575	6.12681	13.19972	0.04348
24	576	13824	4.89898	2.88447	15.49193	6.21435	13.38831	0.04167
25	625	15625	5.00000	2.92399	15.81139	6.29949	13.57173	0.04000
26	676	17576	5.09902	2.96246	16.12452	6.38239	13.75033	0.03846
27	729	19683	5.19615	3.00000	16.43168	6.46318	13.92440	0.03704
28	784	21952	5.29150	3.03656	16.73320	6.54201	14.09422	0.03571
29	841	24389	5.38516	3.07228	17.02939	6.61898	14.26005	0.03448
30	900	27000	5.47723	3.10720	17.32051	6.69420	14.42211	0.03333
31	961	29791	5.56776	3.14134	17.60682	6.76777	14.58061	0.03226
32	1024	32768	5.65685	3.17477	17.88854	6.83977	14.73573	0.03125
33	1089	35937	5.74456	3.20750	18.16590	6.91029	14.88765	0.03030
34	1156	39304	5.83095	3.23957	18.43909	6.97940	15.03654	0.02941
35	1225	42875	5.91608	3.27103	18.70829	7.04716	15.18253	0.02857
36	1296	46656	6.00000	3.30189	18.97367	7.11365	15.32577	0.02778
37	1369	50653	6.08276	3.33218	19.23538	7.17891	15.46638	0.02703
38	1444	54872	6.16441	3.36193	19.49359	7.24301	15.60448	0.02632
39	1521	59319	6.24500	3.39117	19.74842	7.30600	15.74018	0.02564
40	1600	64000	6.32456	3.41991	20.00000	7.36792	15.87357	0.02500
41	1681	68921	6.40312	3.44817	20.24846	7.42881	16.00476	0.02439
42	1764	74088	6.48074	3.47598	20.49390	7.48872	16.13384	0.02381
43	1849	79507	6.55744	3.50335	20.73644	7.54769	16.26088	0.02326
44	1936	85184	6.63325	3.53030	20.97618	7.60575	16.38597	0.02273
45	2025	91125	6.70820	3.55685	21.21320	7.66294	16.50917	0.02222
46	2116	97336	6.78233	3.58300	21.44761	7.71928	16.63057	0.02174
47	2209	103823	6.85565	3.60878	21.67948	7.77482	16.75021	0.02128
48	2304	110592	6.92820	3.63419	21.90890	7.82957	16.86818	0.02083
49	2401	117649	7.00000	3.65926	22.13594	7.88357	16.98451	0.02041
50	2500	125000	7.07107	3.68398	22.36068	7.93684	17.09927	0.02000
51	2601	132651	7.14143	3.70838	22.58318	7.98940	17.21252	0.01961
52	2704	140608	7.21110	3.73246	22.80351	8.04128	17.32429	0.01923
53	2809	148877	7.28011	3.75624	23.02173	8.09250	17.43464	0.01887
54	2916	157464	7.34847	3.77971	23.23790	8.14308	17.54360	0.01852
55	3025	166375	7.41620	3.80290	23.45208	8.19304	17.65123	0.01818
56	3136	175616	7.48331	3.82581	23.66432	8.24240	17.75757	0.01786
57	3249	185193	7.54983	3.84845	23.87467	8.29117	17.86264	0.01754
58	3364	195112	7.61577	3.87082	24.08319	8.33937	17.96650	0.01724
59	3481	205379	7.68115	3.89294	24.28992	8.38703	18.06917	0.01695
60	3600	216000	7.74597	3.91481	24.49490	8.43415	18.17068	0.01667

Table 4 – *continued*

n	n^2	n^3	$\sqrt{n}$	$\sqrt[3]{n}$	$\sqrt{10n}$	$\sqrt[3]{10n}$	$\sqrt[3]{100n}$	$1/n$
61	3721	226981	7.81025	3.93644	24.69818	8.48074	18.27107	0.01639
62	3844	238328	7.87401	3.95784	24.89980	8.52684	18.37037	0.01613
63	3969	250047	7.93725	3.97900	25.09980	8.57243	18.46861	0.01587
64	4096	262144	8.00000	4.00000	25.29822	8.61755	18.56581	0.01563
65	4225	274625	8.06226	4.02067	25.49510	8.66220	18.66201	0.01538
66	4356	287496	8.12404	4.04118	25.69047	8.70640	18.75722	0.01515
67	4489	300763	8.18535	4.06149	25.88436	8.75015	18.85148	0.01493
68	4624	314432	8.24621	4.08160	26.07681	8.79347	18.94481	0.01471
69	4761	328509	8.30662	4.10151	26.26785	8.83636	19.03722	0.01449
70	4900	343000	8.36660	4.12123	26.45751	8.87885	19.12875	0.01429
71	5041	357911	8.42615	4.14076	26.64583	8.92093	19.21941	0.01408
72	5184	373248	8.48528	4.16011	26.83282	8.96261	19.30922	0.01389
73	5329	389017	8.54400	4.17928	27.01851	9.00392	19.39820	0.01370
74	5476	405224	8.60233	4.19828	27.20294	9.04484	19.48637	0.01351
75	5625	421875	8.66025	4.21710	27.38613	9.08540	19.57376	0.01333
76	5776	438976	8.71780	4.23576	27.56810	9.12560	19.66037	0.01316
77	5929	456533	8.77496	4.25426	27.74887	9.16545	19.74622	0.01299
78	6084	474552	8.83176	4.27260	27.92848	9.20496	19.83133	0.01282
79	6241	493039	8.88819	4.29078	28.10694	9.24413	19.91572	0.01266
80	6400	512000	8.94427	4.30881	28.28427	9.28297	19.99940	0.01250
81	6561	531441	9.00000	4.32669	28.46050	9.32149	20.08239	0.01235
82	6724	551368	9.05539	4.34442	28.63564	9.35969	20.16469	0.01220
83	6889	571787	9.11043	4.36201	28.80972	9.39759	20.24633	0.01205
84	7056	592704	9.16515	4.37945	28.98275	9.43518	20.32731	0.01190
85	7225	614125	9.21954	4.39676	29.15476	9.47247	20.40766	0.01176
86	7396	636056	9.27362	4.41394	29.32576	9.50947	20.48738	0.01163
87	7569	658503	9.32738	4.43098	29.49576	9.54619	20.56648	0.01149
88	7744	681472	9.38083	4.44789	29.66479	9.58262	20.64498	0.01136
89	7921	704969	9.43398	4.46468	29.83287	9.61878	20.72288	0.01124
90	8100	729000	9.48683	4.48134	30.00000	9.65467	20.80021	0.01111
91	8281	753571	9.53939	4.49787	30.16621	9.69030	20.87696	0.01099
92	8464	778688	9.59166	4.51429	30.33150	9.72567	20.95315	0.01087
93	8649	804357	9.64365	4.53059	30.49590	9.76078	21.02880	0.01075
94	8836	830584	9.69536	4.54677	30.65942	9.79564	21.10390	0.01064
95	9025	857375	9.74679	4.56283	30.82207	9.83025	21.17847	0.01053
96	9216	884736	9.79796	4.57879	30.98387	9.86462	21.25252	0.01042
97	9409	912673	9.84886	4.59463	31.14482	9.89876	21.32606	0.01031
98	9604	941192	9.89949	4.61037	31.30495	9.93266	21.39909	0.01020
99	9801	970299	9.94987	4.62599	31.46427	9.96633	21.47163	0.01010
100	10000	1000000	10.00000	4.64152	31.62278	9.99977	21.54369	0.01000

Note: Roots of numbers other than those given directly may be found using the following relations:

$$\sqrt{100n} = 10\sqrt{n};\ \sqrt{1000n} = 10\sqrt{10n};\ \sqrt{n/10} = \frac{1}{10}\sqrt{10n};\ \sqrt{n/100} = \frac{1}{10}\sqrt{n};\ \sqrt[3]{1000n} = 10\sqrt[3]{n};\ \sqrt[3]{\frac{n}{10}} = \sqrt[3]{100n}.$$

Table 5. Factorials

The product $n \times (n-1) \times (n-2) \times \cdots \times 2 \times 1$ is called factorial n, expressed as $n!$ **Example:** factorial 5 = 5! = 5 x 4 x 3 x 2 x 1 = 120.

n	$n!$	n	$n!$
0	1	11	39 916 800
1	1	12	479 001 600
2	2	13	6 227 020 800
3	6	14	87 178 291 200
4	24	15	1 307 674 368 000
5	120	16	20 922 789 888 000
6	720	17	355 687 428 096 000
7	5 040	18	6 402 373 705 728 000
8	40 320	19	121 645 100 408 832 000
9	362 880	20	2 432 902 008 176 640 000
10	3 628 800		

Table 6. Logarithms of factorials

Example: $\log_{10} 9! = 5.55976$

n	$\log n!$	n	$\log n!$	n	$\log n!$	n	$\log n!$
1	0.00000	26	26.60562	51	66.19065	76	111.27543
2	0.30103	27	28.03698	52	67.90665	77	113.16192
3	0.77815	28	29.48414	53	69.63092	78	115.05401
4	1.38021	29	30.94654	54	71.36332	79	116.95164
5	2.07918	30	32.42366	55	73.10368	80	118.85473
6	2.85733	31	33.91502	56	74.85187	81	120.76321
7	3.70243	32	35.42017	57	76.60774	82	122.67703
8	4.60552	33	36.93869	58	78.37117	83	124.59610
9	5.55976	34	38.47016	59	80.14202	84	126.52038
10	6.55976	35	40.01423	60	81.92017	85	128.44980
11	7.60116	36	41.57054	61	83.70550	86	130.38430
12	8.68034	37	43.13874	62	85.49790	87	132.32382
13	9.79428	38	44.71852	63	87.29724	88	134.26830
14	10.94041	39	46.30959	64	89.10342	89	136.21769
15	12.11650	40	47.91165	65	90.91633	90	138.17194
16	13.32062	41	49.52443	66	92.73587	91	140.13098
17	14.55107	42	51.14768	67	94.56195	92	142.09477
18	15.80634	43	52.78115	68	96.39446	93	144.06325
19	17.08509	44	54.42460	69	98.23331	94	146.03638
20	18.38612	45	56.07781	70	100.07841	95	148.01410
21	19.70834	46	57.74057	71	101.92966	96	149.99637
22	21.05077	47	59.41267	72	103.78700	97	151.98314
23	22.41249	48	61.09391	73	105.65032	98	153.97437
24	23.79271	49	62.78410	74	107.51955	99	155.97000
25	25.19065	50	64.48307	75	109.39461	100	157.97000

Table 7. Binomial coefficients

$^{n}C_x = \binom{n}{x} = \dfrac{n!}{x!(n-x)!}$. Example: $^{15}C_4 = 1365$.

n	$\binom{n}{0}$	$\binom{n}{1}$	$\binom{n}{2}$	$\binom{n}{3}$	$\binom{n}{4}$	$\binom{n}{5}$	$\binom{n}{6}$	$\binom{n}{7}$	$\binom{n}{8}$	$\binom{n}{9}$	$\binom{n}{10}$
0	1										
1	1	1									
2	1	2	1								
3	1	3	3	1							
4	1	4	6	4	1						
5	1	5	10	10	5	1					
6	1	6	15	20	15	6	1				
7	1	7	21	35	35	21	7	1			
8	1	8	28	56	70	56	28	8	1		
9	1	9	36	84	126	126	84	36	9	1	
10	1	10	45	120	210	252	210	120	45	10	1
11	1	11	55	165	330	462	462	330	165	55	11
12	1	12	66	220	495	792	924	792	495	220	66
13	1	13	78	286	715	1287	1716	1716	1287	715	286
14	1	14	91	364	1001	2002	3003	3432	3003	2002	1001
15	1	15	105	455	1365	3003	5005	6435	6435	5005	3003
16	1	16	120	560	1820	4368	8008	11440	12870	11440	8008
17	1	17	136	680	2380	6188	12376	19448	24310	24310	19448
18	1	18	153	816	3060	8568	18564	31824	43758	48620	43758
19	1	19	171	969	3876	11628	27132	50388	75582	92378	92378
20	1	20	190	1140	4845	15504	38760	77520	125970	167960	184756

For $x > 10$ use the relation: $\binom{n}{x} = \binom{n}{n-x}$. Example: $\binom{20}{14} = \binom{20}{6} = 38760$.

Section 2

Statistical tables

(a) Probability distributions

Table 8. Binomial distribution

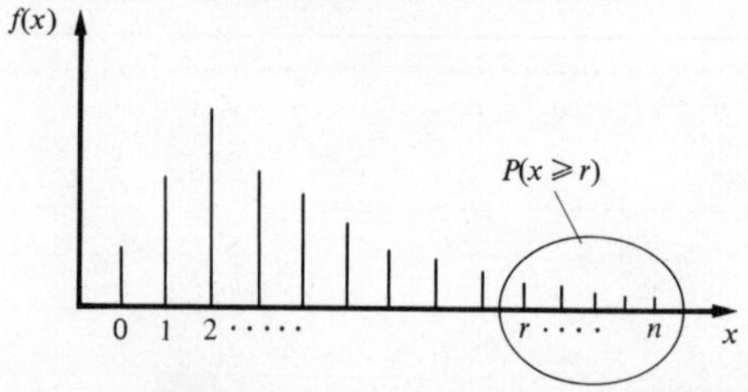

Cumulative probabilities in the right-hand tail, i.e. $P(x \geq r) = \sum_{x=r}^{n} {}^nC_x p^x (1-p)^{n-x}$, for different values of n, r and p.
Example: For $p = 0.2$ and $n = 5$, $P(x \geq 3) = 0.0579$, and $P(x = 3) = P(x \geq 3) - P(x \geq 4) = 0.0579 - 0.0067 = 0.0512$.
If $p \geq 0.5$, $P(x \geq r) = 1 - P(x \geq n - r + 1)$ when p and $(1 - p)$ are interchanged, e.g. for $p = 0.8$ and $n = 5$, $P(x \geq 2) = 1 - P(x \geq 5 - 2 + 1 = 4)$ when $p = 0.2$ and $n = 5$, i.e. $1 - 0.0067 = 0.9933$.

n	r \ p	0.05	0.10	0.15	0.20	0.25	0.30	0.35	0.40	0.45	0.50
2	0	1.0000	1.0000	1.0000	1.0000	1.0000	1.0000	1.0000	1.0000	1.0000	1.0000
	1	0.0975	0.1900	0.2775	0.3600	0.4375	0.5100	0.5775	0.6400	0.6975	0.7500
	2	0.0025	0.0100	0.0225	0.0400	0.0625	0.0900	0.1225	0.1600	0.2025	0.2500
3	0	1.0000	1.0000	1.0000	1.0000	1.0000	1.0000	1.0000	1.0000	1.0000	1.0000
	1	0.1426	0.2710	0.3859	0.4880	0.5781	0.6570	0.7254	0.7840	0.8336	0.8750
	2	0.0073	0.0280	0.0608	0.1040	0.1562	0.2160	0.2818	0.3520	0.4253	0.5000
	3	0.0001	0.0010	0.0034	0.0080	0.0156	0.0270	0.0429	0.0640	0.0911	0.1250
4	0	1.0000	1.0000	1.0000	1.0000	1.0000	1.0000	1.0000	1.0000	1.0000	1.0000
	1	0.1855	0.3439	0.4780	0.5904	0.6836	0.7599	0.8215	0.8704	0.9085	0.9375
	2	0.0140	0.0523	0.1095	0.1808	0.2617	0.3483	0.4370	0.5248	0.6090	0.6875
	3	0.0005	0.0037	0.0120	0.0272	0.0508	0.0837	0.1265	0.1792	0.2415	0.3125
	4		0.0001	0.0005	0.0016	0.0039	0.0081	0.0150	0.0256	0.0410	0.0625
5	0	1.0000	1.0000	1.0000	1.0000	1.0000	1.0000	1.0000	1.0000	1.0000	1.0000
	1	0.2262	0.4095	0.5563	0.6723	0.7627	0.8319	0.8840	0.9222	0.9497	0.9688
	2	0.0226	0.0815	0.1648	0.2627	0.3672	0.4718	0.5716	0.6630	0.7438	0.8125
	3	0.0012	0.0086	0.0266	0.0579	0.1035	0.1631	0.2352	0.3174	0.4069	0.5000
	4		0.0005	0.0022	0.0067	0.0156	0.0308	0.0540	0.0870	0.1312	0.1875
	5			0.0001	0.0003	0.0010	0.0024	0.0053	0.0102	0.0185	0.0313
6	0	1.0000	1.0000	1.0000	1.0000	1.0000	1.0000	1.0000	1.0000	1.0000	1.0000
	1	0.2649	0.4686	0.6229	0.7379	0.8220	0.8824	0.9246	0.9533	0.9723	0.9844
	2	0.0328	0.1143	0.2235	0.3446	0.4661	0.5798	0.6809	0.7667	0.8364	0.8906
	3	0.0022	0.0159	0.0473	0.0989	0.1694	0.2557	0.3529	0.4557	0.5585	0.6563
	4	0.0001	0.0013	0.0059	0.0170	0.0376	0.0705	0.1174	0.1792	0.2553	0.3438
	5		0.0001	0.0004	0.0016	0.0046	0.0109	0.0223	0.0410	0.0692	0.1094
	6				0.0001	0.0002	0.0007	0.0018	0.0041	0.0083	0.0156
7	0	1.0000	1.0000	1.0000	1.0000	1.0000	1.0000	1.0000	1.0000	1.0000	1.0000
	1	0.3017	0.5217	0.6794	0.7903	0.8665	0.9176	0.9510	0.9720	0.9848	0.9922
	2	0.0444	0.1497	0.2834	0.4233	0.5551	0.6706	0.7662	0.8414	0.8976	0.9375
	3	0.0038	0.0257	0.0738	0.1480	0.2436	0.3529	0.4677	0.5801	0.6836	0.7734
	4	0.0002	0.0027	0.0121	0.0333	0.0706	0.1260	0.1998	0.2898	0.3917	0.5000
	5		0.0002	0.0012	0.0047	0.0129	0.0288	0.0556	0.0963	0.1529	0.2266
	6			0.0001	0.0004	0.0013	0.0038	0.0090	0.0188	0.0357	0.0625
	7					0.0001	0.0002	0.0006	0.0016	0.0037	0.0078
8	0	1.0000	1.0000	1.0000	1.0000	1.0000	1.0000	1.0000	1.0000	1.0000	1.0000
	1	0.3366	0.5695	0.7275	0.8322	0.8999	0.9424	0.9681	0.9832	0.9916	0.9961
	2	0.0572	0.1869	0.3428	0.4967	0.6329	0.7447	0.8309	0.8936	0.9368	0.9648
	3	0.0058	0.0381	0.1052	0.2031	0.3215	0.4482	0.5722	0.6846	0.7799	0.8555
	4	0.0004	0.0050	0.0214	0.0563	0.1138	0.1941	0.2936	0.4059	0.5230	0.6367
	5		0.0004	0.0029	0.0104	0.0273	0.0580	0.1061	0.1737	0.2604	0.3633
	6			0.0002	0.0012	0.0042	0.0113	0.0253	0.0498	0.0885	0.1445
	7				0.0001	0.0004	0.0013	0.0036	0.0085	0.0181	0.0352
	8						0.0001	0.0002	0.0007	0.0017	0.0039

Table 8 – *continued*

n	r \ p	0.05	0.10	0.15	0.20	0.25	0.30	0.35	0.40	0.45	0.50
9	0	1.0000	1.0000	1.0000	1.0000	1.0000	1.0000	1.0000	1.0000	1.0000	1.0000
	1	0.3698	0.6126	0.7684	0.8658	0.9249	0.9596	0.9793	0.9899	0.9954	0.9980
	2	0.0712	0.2252	0.4005	0.5638	0.6997	0.8040	0.8789	0.9295	0.9615	0.9805
	3	0.0084	0.0530	0.1409	0.2618	0.3993	0.5372	0.6627	0.7682	0.8505	0.9102
	4	0.0006	0.0083	0.0339	0.0856	0.1657	0.2703	0.3911	0.5174	0.6386	0.7461
	5		0.0009	0.0056	0.0196	0.0489	0.0988	0.1717	0.2666	0.3786	0.5000
	6		0.0001	0.0006	0.0031	0.0100	0.0253	0.0536	0.0994	0.1658	0.2539
	7				0.0003	0.0013	0.0043	0.0112	0.0250	0.0498	0.0898
	8					0.0001	0.0004	0.0014	0.0038	0.0091	0.0195
	9							0.0001	0.0003	0.0008	0.0020
10	0	1.0000	1.0000	1.0000	1.0000	1.0000	1.0000	1.0000	1.0000	1.0000	1.0000
	1	0.4013	0.6513	0.8031	0.8926	0.9437	0.9718	0.9865	0.9940	0.9975	0.9990
	2	0.0861	0.2639	0.4557	0.6242	0.7560	0.8507	0.9140	0.9536	0.9767	0.9893
	3	0.0115	0.0702	0.1798	0.3222	0.4744	0.6172	0.7384	0.8327	0.9004	0.9453
	4	0.0010	0.0128	0.0500	0.1209	0.2241	0.3504	0.4862	0.6177	0.7340	0.8281
	5	0.0001	0.0016	0.0099	0.0328	0.0781	0.1503	0.2485	0.3669	0.4956	0.6230
	6		0.0001	0.0014	0.0064	0.0197	0.0473	0.0949	0.1662	0.2616	0.3770
	7			0.0001	0.0009	0.0035	0.0106	0.0260	0.0548	0.1020	0.1719
	8				0.0001	0.0004	0.0016	0.0048	0.0123	0.0274	0.0547
	9						0.0001	0.0005	0.0017	0.0045	0.0107
	10								0.0001	0.0003	0.0010
12	0	1.0000	1.0000	1.0000	1.0000	1.0000	1.0000	1.0000	1.0000	1.0000	1.0000
	1	0.4596	0.7176	0.8578	0.9313	0.9683	0.9862	0.9943	0.9978	0.9992	0.9998
	2	0.1184	0.3410	0.5565	0.7251	0.8416	0.9150	0.9576	0.9804	0.9917	0.9968
	3	0.0196	0.1109	0.2642	0.4417	0.6093	0.7472	0.8487	0.9166	0.9579	0.9807
	4	0.0022	0.0256	0.0922	0.2054	0.3512	0.5075	0.6533	0.7747	0.8655	0.9270
	5	0.0002	0.0043	0.0239	0.0726	0.1576	0.2763	0.4167	0.5618	0.6956	0.8062
	6		0.0005	0.0046	0.0194	0.0544	0.1178	0.2127	0.3348	0.4731	0.6128
	7		0.0001	0.0007	0.0039	0.0143	0.0386	0.0846	0.1582	0.2607	0.3872
	8			0.0001	0.0006	0.0028	0.0095	0.0255	0.0573	0.1117	0.1938
	9				0.0001	0.0004	0.0017	0.0056	0.0153	0.0356	0.0730
	10						0.0002	0.0008	0.0028	0.0079	0.0193
	11							0.0001	0.0003	0.0011	0.0032
	12									0.0001	0.0002
14	0	1.0000	1.0000	1.0000	1.0000	1.0000	1.0000	1.0000	1.0000	1.0000	1.0000
	1	0.5123	0.7712	0.8972	0.9560	0.9822	0.9932	0.9976	0.9992	0.9998	0.9999
	2	0.1530	0.4154	0.6433	0.8021	0.8990	0.9525	0.9795	0.9919	0.9971	0.9991
	3	0.0301	0.1584	0.3521	0.5519	0.7189	0.8392	0.9161	0.9602	0.9830	0.9935
	4	0.0042	0.0441	0.1465	0.3018	0.4787	0.6448	0.7795	0.8757	0.9368	0.9713
	5	0.0004	0.0092	0.0467	0.1298	0.2585	0.4158	0.5773	0.7207	0.8328	0.9102
	6		0.0015	0.0115	0.0439	0.1117	0.2195	0.3595	0.5141	0.6627	0.7880
	7		0.0002	0.0022	0.0116	0.0383	0.0933	0.1836	0.3075	0.4539	0.6047
	8			0.0003	0.0024	0.0103	0.0315	0.0753	0.1501	0.2586	0.3953
	9				0.0004	0.0022	0.0083	0.0243	0.0583	0.1189	0.2120
	10					0.0003	0.0017	0.0060	0.0175	0.0426	0.0898
	11						0.0002	0.0011	0.0039	0.0114	0.0287
	12							0.0001	0.0006	0.0022	0.0065
	13								0.0001	0.0003	0.0009
	14										0.0001

Table 8 – *continued*

n	r \ p	0.05	0.10	0.15	0.20	0.25	0.30	0.35	0.40	0.45	0.50
16	0	1.0000	1.0000	1.0000	1.0000	1.0000	1.0000	1.0000	1.0000	1.0000	1.0000
	1	0.5599	0.8147	0.9257	0.9719	0.9900	0.9967	0.9990	0.9997	0.9999	1.0000
	2	0.1892	0.4853	0.7161	0.8593	0.9365	0.9739	0.9902	0.9967	0.9990	0.9997
	3	0.0429	0.2108	0.4386	0.6482	0.8029	0.9006	0.9549	0.9817	0.9934	0.9979
	4	0.0070	0.0684	0.2101	0.4019	0.5950	0.7541	0.8661	0.9349	0.9719	0.9894
	5	0.0009	0.0170	0.0791	0.2018	0.3698	0.5501	0.7108	0.8334	0.9147	0.9616
	6	0.0001	0.0033	0.0235	0.0817	0.1897	0.3402	0.5100	0.6712	0.8024	0.8949
	7		0.0005	0.0056	0.0267	0.0796	0.1753	0.3119	0.4738	0.6340	0.7728
	8		0.0001	0.0011	0.0070	0.0271	0.0744	0.1594	0.2839	0.4371	0.5982
	9			0.0002	0.0015	0.0075	0.0257	0.0671	0.1423	0.2559	0.4018
	10				0.0002	0.0016	0.0071	0.0229	0.0583	0.1241	0.2272
	11					0.0003	0.0016	0.0062	0.0191	0.0486	0.1051
	12						0.0003	0.0013	0.0049	0.0149	0.0384
	13							0.0002	0.0009	0.0035	0.0106
	14								0.0001	0.0006	0.0021
	15									0.0001	0.0003
18	0	1.0000	1.0000	1.0000	1.0000	1.0000	1.0000	1.0000	1.0000	1.0000	1.0000
	1	0.6028	0.8499	0.9464	0.9820	0.9944	0.9984	0.9996	0.9999	1.0000	1.0000
	2	0.2265	0.5497	0.7759	0.9009	0.9605	0.9858	0.9954	0.9987	0.9997	0.9999
	3	0.0581	0.2662	0.5203	0.7287	0.8647	0.9400	0.9764	0.9918	0.9975	0.9993
	4	0.0109	0.0982	0.2798	0.4990	0.6943	0.8354	0.9217	0.9672	0.9880	0.9962
	5	0.0015	0.0282	0.1206	0.2836	0.4813	0.6673	0.8114	0.9058	0.9589	0.9846
	6	0.0002	0.0064	0.0419	0.1329	0.2825	0.4656	0.6450	0.7912	0.8923	0.9519
	7		0.0012	0.0118	0.0513	0.1390	0.2783	0.4509	0.6257	0.7742	0.8811
	8		0.0002	0.0027	0.0163	0.0569	0.1407	0.2717	0.4366	0.6085	0.7597
	9			0.0005	0.0043	0.0193	0.0596	0.1391	0.2632	0.4222	0.5927
	10			0.0001	0.0009	0.0054	0.0210	0.0597	0.1347	0.2527	0.4073
	11				0.0002	0.0012	0.0061	0.0212	0.0576	0.1280	0.2403
	12					0.0002	0.0014	0.0062	0.0203	0.0537	0.1189
	13						0.0003	0.0014	0.0058	0.0183	0.0481
	14							0.0003	0.0013	0.0049	0.0154
	15								0.0002	0.0010	0.0038
	16									0.0001	0.0007
	17										0.0001
20	0	1.0000	1.0000	1.0000	1.0000	1.0000	1.0000	1.0000	1.0000	1.0000	1.0000
	1	0.6415	0.8784	0.9612	0.9885	0.9968	0.9992	0.9998	1.0000	1.0000	1.0000
	2	0.2642	0.6083	0.8244	0.9308	0.9757	0.9924	0.9979	0.9995	0.9999	1.0000
	3	0.0755	0.3231	0.5951	0.7939	0.9087	0.9645	0.9879	0.9964	0.9991	0.9998
	4	0.0159	0.1330	0.3523	0.5886	0.7748	0.8929	0.9556	0.9840	0.9951	0.9987
	5	0.0026	0.0432	0.1702	0.3704	0.5852	0.7625	0.8818	0.9490	0.9811	0.9941
	6	0.0003	0.0113	0.0673	0.1958	0.3828	0.5836	0.7546	0.8744	0.9447	0.9793
	7		0.0024	0.0219	0.0867	0.2142	0.3920	0.5834	0.7500	0.8701	0.9423
	8		0.0004	0.0059	0.0321	0.1018	0.2277	0.3990	0.5841	0.7480	0.8684
	9		0.0001	0.0013	0.0100	0.0409	0.1133	0.2376	0.4044	0.5857	0.7483
	10			0.0002	0.0026	0.0139	0.0480	0.1218	0.2447	0.4086	0.5881
	11				0.0006	0.0039	0.0171	0.0532	0.1275	0.2493	0.4119
	12				0.0001	0.0009	0.0051	0.0196	0.0565	0.1308	0.2517
	13					0.0002	0.0013	0.0060	0.0210	0.0580	0.1316
	14						0.0003	0.0015	0.0065	0.0214	0.0577
	15							0.0003	0.0016	0.0064	0.0207
	16								0.0003	0.0015	0.0059
	17									0.0003	0.0013
	18										0.0002

Approximation: 1. When $p < 0.1$ use the Poisson distribution with $m = np$ (see Table 9).
2. When $0.1 \leqslant p \leqslant 0.9$ and $np > 5$ use Normal distribution with $\mu = np$ and $\sigma = \sqrt{np(1-p)}$ (see Table 11).

Table 9. Poisson distribution

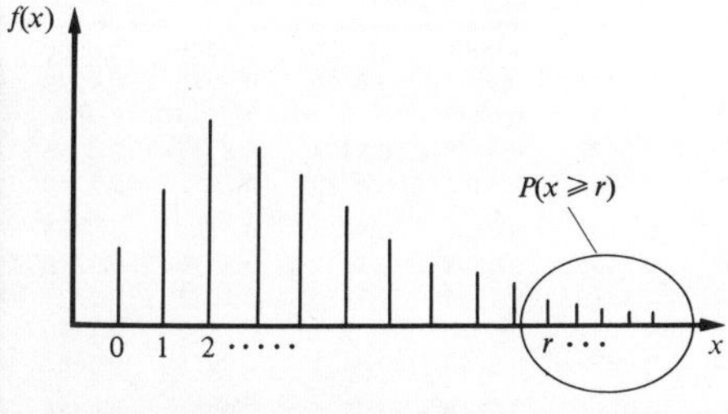

Cumulative probabilities in the right-hand tail, i.e. $P(x \geqslant r) = \sum_{x=r}^{\infty} e^{-m} \frac{m^x}{x!}$, for different values of m and r. **Example:** For m (mean) = 2, $P(x \geqslant r = 3) = 0.3233$ and $P(x = 3) = P(x \geqslant 3) - P(x \geqslant 4) = 0.3233 - 0.1429 = 0.1804$.

r \ m	0.1	0.2	0.3	0.4	0.5	0.6	0.7	0.8	0.9	1.0
0	1.0000	1.0000	1.0000	1.0000	1.0000	1.0000	1.0000	1.0000	1.0000	1.0000
1	0.0952	0.1813	0.2592	0.3297	0.3935	0.4512	0.5034	0.5507	0.5934	0.6321
2	0.0047	0.0175	0.0369	0.0616	0.0902	0.1219	0.1558	0.1912	0.2275	0.2642
3	0.0002	0.0011	0.0036	0.0079	0.0144	0.0231	0.0341	0.0474	0.0629	0.0803
4		0.0001	0.0003	0.0008	0.0018	0.0034	0.0058	0.0091	0.0135	0.0190
5				0.0001	0.0002	0.0004	0.0008	0.0014	0.0023	0.0037
6							0.0001	0.0002	0.0003	0.0006
7										0.0001

r \ m	1.1	1.2	1.3	1.4	1.5	1.6	1.7	1.8	1.9	2.0
0	1.0000	1.0000	1.0000	1.0000	1.0000	1.0000	1.0000	1.0000	1.0000	1.0000
1	0.6671	0.6988	0.7275	0.7534	0.7769	0.7981	0.8173	0.8347	0.8504	0.8647
2	0.3010	0.3374	0.3732	0.4082	0.4422	0.4751	0.5068	0.5372	0.5663	0.5940
3	0.0996	0.1205	0.1249	0.1665	0.1912	0.2166	0.2428	0.2694	0.2963	0.3233
4	0.0257	0.0338	0.0431	0.0537	0.0656	0.0788	0.0932	0.1087	0.1253	0.1429
5	0.0054	0.0077	0.0107	0.0143	0.0186	0.0237	0.0296	0.0364	0.0441	0.0527
6	0.0010	0.0015	0.0022	0.0032	0.0045	0.0060	0.0080	0.0104	0.0132	0.0166
7	0.0001	0.0003	0.0004	0.0006	0.0009	0.0013	0.0019	0.0026	0.0034	0.0045
8			0.0001	0.0001	0.0002	0.0003	0.0004	0.0006	0.0008	0.0011
9							0.0001	0.0001	0.0002	0.0002

r \ m	2.1	2.2	2.3	2.4	2.5	2.6	2.7	2.8	2.9	3.0
0	1.0000	1.0000	1.0000	1.0000	1.0000	1.0000	1.0000	1.0000	1.0000	1.0000
1	0.8775	0.8892	0.8997	0.9093	0.9179	0.9257	0.9328	0.9392	0.9450	0.9502
2	0.6204	0.6454	0.6691	0.6916	0.7127	0.7326	0.7513	0.7689	0.7854	0.8009
3	0.3504	0.3773	0.4040	0.4303	0.4562	0.4816	0.5064	0.5305	0.5540	0.5768
4	0.1614	0.1806	0.2007	0.2213	0.2424	0.2640	0.2859	0.3081	0.3304	0.3528
5	0.0621	0.0725	0.0838	0.0959	0.1088	0.1226	0.1371	0.1523	0.1682	0.1847
6	0.0204	0.0249	0.0300	0.0357	0.0420	0.0490	0.0567	0.0651	0.0742	0.0839
7	0.0059	0.0075	0.0094	0.0116	0.0142	0.0172	0.0206	0.0244	0.0287	0.0335
8	0.0015	0.0020	0.0026	0.0033	0.0042	0.0053	0.0066	0.0081	0.0099	0.0119
9	0.0003	0.0005	0.0006	0.0009	0.0011	0.0015	0.0019	0.0024	0.0031	0.0038
10	0.0001	0.0001	0.0001	0.0002	0.0003	0.0004	0.0005	0.0007	0.0009	0.0011
11					0.0001	0.0001	0.0001	0.0002	0.0002	0.0003
12									0.0001	0.0001

Table 9 – *continued*

r \ m	3.2	3.4	3.6	3.8	4.0	4.2	4.4	4.6	4.8	5.0
0	1.0000	1.0000	1.0000	1.0000	1.0000	1.0000	1.0000	1.0000	1.0000	1.0000
1	0.9592	0.9666	0.9727	0.9776	0.9817	0.9850	0.9877	0.9899	0.9918	0.9933
2	0.8288	0.8532	0.8743	0.8926	0.9084	0.9220	0.9337	0.9437	0.9523	0.9596
3	0.6201	0.6603	0.6973	0.7311	0.7619	0.7898	0.8149	0.8374	0.8575	0.8753
4	0.3975	0.4416	0.4848	0.5265	0.5665	0.6046	0.6406	0.6743	0.7058	0.7350
5	0.2194	0.2558	0.2936	0.3322	0.3712	0.4102	0.4488	0.4868	0.5237	0.5595
6	0.1054	0.1295	0.1559	0.1844	0.2149	0.2469	0.2801	0.3142	0.3490	0.3840
7	0.0446	0.0579	0.0733	0.0909	0.1107	0.1325	0.1564	0.1820	0.2092	0.2378
8	0.0168	0.0231	0.0308	0.0401	0.0511	0.0639	0.0786	0.0951	0.1133	0.1334
9	0.0057	0.0083	0.0117	0.0160	0.0214	0.0279	0.0358	0.0451	0.0558	0.0681
10	0.0018	0.0027	0.0040	0.0058	0.0081	0.0111	0.0149	0.0195	0.0251	0.0318
11	0.0005	0.0008	0.0013	0.0019	0.0028	0.0041	0.0057	0.0078	0.0104	0.0137
12	0.0001	0.0002	0.0004	0.0006	0.0009	0.0014	0.0020	0.0029	0.0040	0.0055
13		0.0001	0.0001	0.0002	0.0003	0.0004	0.0007	0.0010	0.0014	0.0020
14					0.0001	0.0001	0.0002	0.0003	0.0005	0.0007
15							0.0001	0.0001	0.0001	0.0002
16										0.0001

r \ m	5.5	6.0	6.5	7.0	7.5	8.0	8.5	9.0	9.5	10.0
0	1.0000	1.0000	1.0000	1.0000	1.0000	1.0000	1.0000	1.0000	1.0000	1.0000
1	0.9959	0.9975	0.9985	0.9991	0.9994	0.9997	0.9998	0.9999	0.9999	1.0000
2	0.9734	0.9826	0.9887	0.9927	0.9953	0.9970	0.9981	0.9988	0.9992	0.9995
3	0.9116	0.9380	0.9570	0.9704	0.9797	0.9862	0.9907	0.9938	0.9958	0.9972
4	0.7983	0.8488	0.8882	0.9182	0.9409	0.9576	0.9699	0.9788	0.9851	0.9897
5	0.6425	0.7149	0.7763	0.8270	0.8679	0.9004	0.9256	0.9450	0.9597	0.9707
6	0.4711	0.5543	0.6310	0.6993	0.7586	0.8088	0.8504	0.8843	0.9115	0.9329
7	0.3140	0.3937	0.4735	0.5503	0.6218	0.6866	0.7438	0.7932	0.8351	0.8699
8	0.1905	0.2560	0.3272	0.4013	0.4754	0.5470	0.6144	0.6761	0.7313	0.7798
9	0.1056	0.1528	0.2084	0.2709	0.3380	0.4075	0.4769	0.5443	0.6082	0.6672
10	0.0538	0.0839	0.1226	0.1695	0.2236	0.2834	0.3470	0.4126	0.4782	0.5421
11	0.0253	0.0426	0.0668	0.0985	0.1378	0.1841	0.2366	0.2940	0.3547	0.4170
12	0.0110	0.0201	0.0339	0.0533	0.0792	0.1119	0.1513	0.1970	0.2480	0.3032
13	0.0045	0.0088	0.0160	0.0270	0.0427	0.0638	0.0909	0.1242	0.1636	0.2084
14	0.0017	0.0036	0.0071	0.0128	0.0216	0.0342	0.0514	0.0739	0.1019	0.1355
15	0.0006	0.0014	0.0030	0.0057	0.0103	0.0173	0.0274	0.0415	0.0600	0.0835
16	0.0002	0.0005	0.0012	0.0024	0.0046	0.0082	0.0138	0.0220	0.0335	0.0487
17	0.0001	0.0002	0.0004	0.0010	0.0020	0.0037	0.0066	0.0111	0.0177	0.0270
18		0.0001	0.0002	0.0004	0.0008	0.0016	0.0030	0.0053	0.0089	0.0143
19			0.0001	0.0001	0.0003	0.0007	0.0013	0.0024	0.0043	0.0072
20					0.0001	0.0003	0.0005	0.0011	0.0020	0.0035
21						0.0001	0.0002	0.0004	0.0009	0.0016
22							0.0001	0.0002	0.0004	0.0007
23								0.0001	0.0001	0.0003
24									0.0001	0.0001

Table 10. Negative exponential distribution, e^{-x}

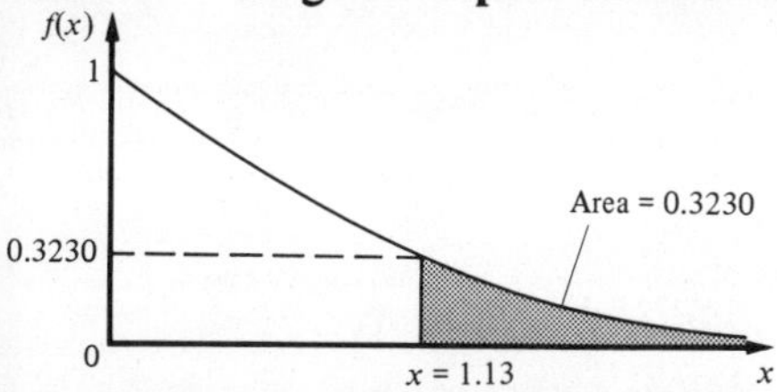

Values of $f(x) = e^{-x}$ for different values of x. **Example:** $f(x = 1.13) = e^{-1.13} = 0.3230$. Area under $f(x)$ beyond $x = 1.13$ is also given by 0.3230. Area under $f(x) = \theta e^{-\theta x}$ beyond x is given by $e^{-\theta x}$. **Example:** For $\theta = \frac{1}{5}$ area beyond $x = 12$ is given by $e^{-12/5} = e^{-2.4} = 0.0907$.

$x \rightarrow$ / $\downarrow$	0.00	0.01	0.02	0.03	0.04	0.05	0.06	0.07	0.08	0.09
0	1.0000	0.9900	0.9802	0.9704	0.9608	0.9512	0.9418	0.9324	0.9231	0.9139
0.1	0.9048	0.8958	0.8869	0.8781	0.8694	0.8607	0.8521	0.8437	0.8353	0.8270
0.2	0.8187	0.8106	0.8025	0.7945	0.7866	0.7788	0.7711	0.7634	0.7558	0.7483
0.3	0.7408	0.7334	0.7261	0.7189	0.7118	0.7047	0.6977	0.6907	0.6839	0.6771
0.4	0.6703	0.6637	0.6570	0.6505	0.6440	0.6376	0.6313	0.6250	0.6188	0.6126
0.5	0.6065	0.6005	0.5945	0.5886	0.5827	0.5769	0.5712	0.5655	0.5599	0.5543
0.6	0.5488	0.5434	0.5379	0.5326	0.5273	0.5220	0.5169	0.5117	0.5066	0.5016
0.7	0.4966	0.4916	0.4868	0.4819	0.4771	0.4724	0.4677	0.4630	0.4584	0.4538
0.8	0.4493	0.4449	0.4404	0.4360	0.4317	0.4274	0.4232	0.4190	0.4148	0.4107
0.9	0.4066	0.4025	0.3985	0.3946	0.3906	0.3867	0.3829	0.3791	0.3753	0.3716
1.0	0.3679	0.3642	0.3606	0.3570	0.3535	0.3499	0.3465	0.3430	0.3396	0.3362
1.1	0.3329	0.3296	0.3265	0.3230	0.3198	0.3166	0.3135	0.3104	0.3073	0.3042
1.2	0.3012	0.2892	0.2952	0.2923	0.2894	0.2865	0.2837	0.2808	0.2780	0.2753
1.3	0.2725	0.2698	0.2671	0.2645	0.2618	0.2592	0.2567	0.2541	0.2516	0.2491
1.4	0.2466	0.2441	0.2417	0.2393	0.2369	0.2346	0.2322	0.2299	0.2276	0.2254
1.5	0.2231	0.2209	0.2187	0.2165	0.2144	0.2122	0.2101	0.2080	0.2060	0.2039
1.6	0.2019	0.1999	0.1979	0.1959	0.1940	0.1920	0.1901	0.1882	0.1864	0.1845
1.7	0.1827	0.1809	0.1791	0.1773	0.1755	0.1738	0.1720	0.1703	0.1686	0.1670
1.8	0.1653	0.1637	0.1620	0.1604	0.1588	0.1572	0.1557	0.1541	0.1526	0.1511
1.9	0.1496	0.1481	0.1466	0.1451	0.1437	0.1423	0.1409	0.1395	0.1381	0.1367
2.0	0.1353	0.1340	0.1327	0.1313	0.1300	0.1287	0.1275	0.1262	0.1249	0.1237
2.1	0.1225	0.1212	0.1200	0.1188	0.1177	0.1165	0.1153	0.1142	0.1130	0.1119
2.2	0.1108	0.1097	0.1086	0.1075	0.1065	0.1054	0.1044	0.1035	0.1023	0.1013
2.3	0.1003	0.0993	0.0983	0.0973	0.0963	0.0954	0.0944	0.0935	0.0926	0.0916
2.4	0.0907	0.0898	0.0889	0.0880	0.0872	0.0863	0.0854	0.0846	0.0837	0.0829
2.5	0.0821	0.0813	0.0805	0.0797	0.0789	0.0781	0.0773	0.0765	0.0758	0.0750
2.6	0.0743	0.0735	0.0728	0.0721	0.0714	0.0707	0.0699	0.0693	0.0686	0.0679
2.7	0.0672	0.0665	0.0659	0.0652	0.0646	0.0639	0.0633	0.0627	0.0620	0.0614
2.8	0.0608	0.0602	0.0596	0.0590	0.0584	0.0578	0.0573	0.0567	0.0561	0.0556
2.9	0.0550	0.0545	0.0539	0.0534	0.0529	0.0523	0.0518	0.0513	0.0508	0.0503
3.0	0.0498	0.0493	0.0488	0.0483	0.0478	0.0474	0.0469	0.0464	0.0460	0.0455
3.1	0.0450	0.0446	0.0442	0.0437	0.0433	0.0429	0.0424	0.0420	0.0416	0.0412
3.2	0.0408	0.0404	0.0400	0.0396	0.0392	0.0388	0.0384	0.0380	0.0376	0.0373
3.3	0.0369	0.0365	0.0362	0.0358	0.0354	0.0351	0.0347	0.0344	0.0340	0.0337
3.4	0.0334	0.0330	0.0327	0.0324	0.0321	0.0317	0.0314	0.0311	0.0308	0.0305
3.5	0.0302	0.0299	0.0296	0.0293	0.0290	0.0287	0.0284	0.0282	0.0279	0.0276
3.6	0.0273	0.0271	0.0268	0.0265	0.0263	0.0260	0.0257	0.0255	0.0252	0.0250
3.7	0.0247	0.0245	0.0242	0.0240	0.0238	0.0235	0.0233	0.0231	0.0228	0.0226
3.8	0.0224	0.0221	0.0219	0.0217	0.0215	0.0213	0.0211	0.0209	0.0207	0.0204
3.9	0.0202	0.0200	0.0198	0.0196	0.0194	0.0193	0.0191	0.0189	0.0187	0.0185
4.0	0.0183	0.0181	0.0180	0.0178	0.0176	0.0174	0.0172	0.0171	0.0169	0.0167
4.1	0.0166	0.0164	0.0162	0.0161	0.0159	0.0158	0.0156	0.0155	0.0153	0.0151
4.2	0.0150	0.0148	0.0147	0.0146	0.0144	0.0143	0.0141	0.0140	0.0138	0.0137
4.3	0.0136	0.0134	0.0133	0.0132	0.0130	0.0129	0.0128	0.0127	0.0125	0.0124
4.4	0.0123	0.0122	0.0120	0.0119	0.0118	0.0117	0.0116	0.0114	0.0113	0.0112
4.5	0.0111	0.0110	0.0109	0.0108	0.0107	0.0106	0.0105	0.0104	0.0103	0.0102
4.6	0.0101	0.0100	0.0099	0.0098	0.0097	0.0096	0.0095	0.0094	0.0093	0.0092
4.7	0.0091	0.0090	0.0089	0.0088	0.0087	0.0087	0.0086	0.0085	0.0084	0.0083
4.8	0.0082	0.0081	0.0081	0.0080	0.0079	0.0078	0.0078	0.0077	0.0076	0.0075
4.9	0.0074	0.0074	0.0073	0.0072	0.0072	0.0071	0.0070	0.0069	0.0069	0.0068
5.0	0.0067	0.0067	0.0066	0.0065	0.0065	0.0064	0.0063	0.0063	0.0062	0.0062

x	6.0	7.0	8.0	9.0	10.0	12.0	14.0	16.0	18.0	20.0
e^{-x}	0.0^2248	0.0^3912	0.0^3335	0.0^3123	0.0^4454	0.0^5614	0.0^6832	0.0^6113	0.0^7152	0.0^8206

Note: To find e^{-x} for $x > 5.09$ use the rule on exponents, e.g. $e^{-8.25} = e^{-8}e^{-0.25} = 0.0^3335 \times 0.7788 = 0.000335 \times 0.7788 = 0.000261$.

Table 11. Normal distribution (areas)

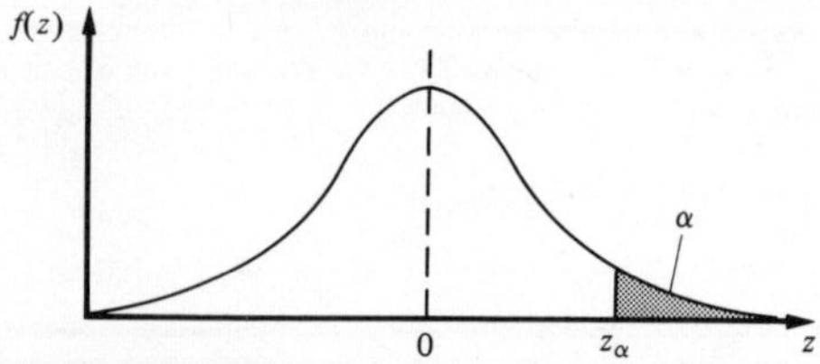

Area (α) in the tail of the standardised Normal curve, $N(0, 1)$, for different values of z. **Example:** Area beyond $z = 1.96$ (or below $z = -1.96$) is $\alpha = 0.02500$. For Normal curve with $\mu = 10$ and $\sigma = 2$, area beyond $x = 12$, say, is the same as area beyond $z = \frac{x-\mu}{\sigma} = \frac{12-10}{2} = 1$, i.e. $\alpha = 0.15866$.

z→ ↓	0.00	0.01	0.02	0.03	0.04	0.05	0.06	0.07	0.08	0.09
0.0	.50000	.49601	.49202	.48803	.48405	.48006	.47608	.47210	.46812	.46414
0.1	.46017	.45620	.45224	.44828	.44433	.44038	.43644	.43251	.42858	.42465
0.2	.42074	.41683	.41294	.40905	.40517	.40129	.39743	.39358	.38974	.38591
0.3	.38209	.37828	.37448	.37070	.36693	.36317	.35942	.35569	.35197	.34827
0.4	.34458	.34090	.33724	.33360	.32997	.32636	.32276	.31918	.31561	.31207
0.5	.30854	.30503	.30153	.29806	.29460	.29116	.28774	.28434	.28096	.27760
0.6	.27425	.27093	.26763	.26435	.26109	.25785	.25463	.25143	.24825	.24510
0.7	.24196	.23885	.23576	.23270	.22965	.22663	.22363	.22065	.21770	.21476
0.8	.21186	.20897	.20611	.20327	.20045	.19766	.19489	.19215	.18943	.18673
0.9	.18406	.18141	.17879	.17619	.17361	.17106	.16853	.16602	.16354	.16109
1.0	.15866	.15625	.15386	.15150	.14917	.14686	.14457	.14231	.14007	.13786
1.1	.13567	.13350	.13136	.12924	.12714	.12507	.12302	.12100	.11900	.11702
1.2	.11507	.11314	.11123	.10935	.10749	.10565	.10383	.10204	.10027	.09853
1.3	.09680	.09510	.09342	.09176	.09012	.08851	.08692	.08534	.08379	.08226
1.4	.08076	.07927	.07780	.07636	.07493	.07353	.07214	.07078	.06944	.06811
1.5	.06681	.06552	.06426	.06301	.06178	.06057	.05938	.05821	.05705	.05592
1.6	.05480	.05370	.05262	.05155	.05050	.04947	.04846	.04746	.04648	.04551
1.7	.04457	.04363	.04272	.04182	.04093	.04006	.03920	.03836	.03754	.03673
1.8	.03593	.03515	.03438	.03362	.03288	.03216	.03144	.03074	.03005	.02938
1.9	.02872	.02807	.02743	.02680	.02619	.02559	.02500	.02442	.02385	.02330
2.0	.02275	.02222	.02169	.02118	.02068	.02018	.01970	.01923	.01876	.01831
2.1	.01786	.01743	.01700	.01659	.01618	.01578	.01539	.01500	.01463	.01426
2.2	.01390	.01355	.01321	.01287	.01254	.01222	.01191	.01160	.01130	.01101
2.3	.01072	.01044	.01017	.00990	.00964	.00939	.00914	.00889	.00866	.00842
2.4	.00820	.00798	.00776	.00755	.00734	.00714	.00695	.00676	.00657	.00639
2.5	.00621	.00604	.00587	.00570	.00554	.00539	.00523	.00509	.00494	.00480
2.6	.00466	.00453	.00440	.00427	.00415	.00403	.00391	.00379	.00368	.00357
2.7	.00347	.00336	.00326	.00317	.00307	.00298	.00289	.00280	.00272	.00263
2.8	.00256	.00248	.00240	.00233	.00226	.00219	.00212	.00205	.00199	.00193
2.9	.00187	.00181	.00175	.00169	.00164	.00159	.00154	.00149	.00144	.00139
3.0	.00135	.00131	.00126	.00122	.00118	.00114	.00111	.00107	.00104	.00100
3.1	.00097	.00094	.00090	.00087	.00085	.00082	.00079	.00076	.00074	.00071
3.2	.00069	.00066	.00064	.00062	.00060	.00058	.00056	.00054	.00052	.00050
3.3	.00048	.00047	.00045	.00043	.00042	.00040	.00039	.00038	.00036	.00035
3.4	.00034	.00032	.00031	.00030	.00029	.00028	.00027	.00026	.00025	.00024
3.5	.00023	.00022	.00022	.00021	.00020	.00019	.00019	.00018	.00017	.00017
3.6	.00016	.00015	.00015	.00014	.00014	.00013	.00013	.00012	.00012	.00011
3.7	.00011	.00010	.00010	.00010	.00009	.00009	.00009	.00008	.00008	.00008
3.8	.00007	.00007	.00007	.00006	.00006	.00006	.00006	.00005	.00005	.00005
3.9	.00005	.00005	.00004	.00004	.00004	.00004	.00004	.00004	.00004	.00003
4.0	.00003	.00003	.00003	.00003	.00003	.00002	.00002	.00002	.00002	.00002

α	0.4	0.25	0.2	0.15	0.1	0.05	0.025	0.01	0.005	0.001
z_α	.2533	.6745	.8416	1.0364	1.2816	1.6449	1.9600	2.3263	2.5758	3.0902

Table 12. Normal distribution (ordinates)

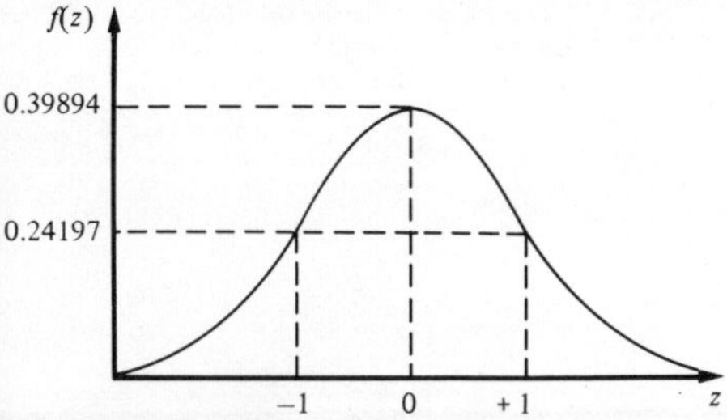

Ordinates of the standardised Normal curve, $N(0, 1)$. **Example:** $f(z = 1) = f(z = -1) = 0.24197$ and $f(z = 0) = 0.39894$. For $N(\mu, \sigma^2)$, $f(x) = f(z = (x - \mu)/\sigma)/\sigma$. **Example:** For $\mu = 10$ and $\sigma = 2$, $f(x = 12) = f(z = (12 - 10)/2)/2 = f(z = 1)/2 = (0.24197)/2 = 0.12099$.

z→ ↓	0.00	0.01	0.02	0.03	0.04	0.05	0.06	0.07	0.08	0.09
0.0	.39894	.39892	.39886	.39876	.39862	.39844	.39822	.39797	.39767	.39733
0.1	.39695	.39654	.39608	.39559	.39505	.39448	.39387	.39322	.39253	.39181
0.2	.39104	.39024	.38940	.38853	.38762	.38667	.38568	.38466	.38361	.38251
0.3	.38139	.38023	.37903	.37780	.37654	.37524	.37391	.37255	.37115	.36973
0.4	.36827	.36678	.36526	.36371	.36213	.36053	.35889	.35723	.35553	.35381
0.5	.35207	.35029	.34849	.34667	.34482	.34294	.34105	.33912	.33718	.33521
0.6	.33322	.33121	.32918	.32713	.32506	.32297	.32086	.31874	.31659	.31443
0.7	.31225	.31006	.30785	.30563	.30339	.30114	.29887	.29659	.29431	.29200
0.8	.28969	.28737	.28504	.28269	.28034	.27798	.27562	.27324	.27086	.26848
0.9	.26609	.26369	.26129	.25888	.25647	.25406	.25164	.24923	.24681	.24439
1.0	.24197	.23955	.23713	.23471	.23230	.22988	.22747	.22506	.22265	.22025
1.1	.21785	.21546	.21307	.21069	.20831	.20594	.20357	.20121	.19886	.19652
1.2	.19419	.19186	.18954	.18724	.18494	.18265	.18037	.17810	.17585	.17360
1.3	.17137	.16915	.16694	.16474	.16256	.16038	.15822	.15608	.15395	.15183
1.4	.14973	.14764	.14556	.14350	.14146	.13943	.13742	.13542	.13344	.13147
1.5	.12952	.12758	.12566	.12376	.12188	.12001	.11816	.11632	.11450	.11270
1.6	.11092	.10915	.10741	.10567	.10396	.10226	.10059	.09893	.09728	.09566
1.7	.09405	.09246	.09089	.08933	.08780	.08628	.08478	.08329	.08183	.08038
1.8	.07895	.07754	.07614	.07477	.07341	.07206	.07074	.06943	.06814	.06687
1.9	.06562	.06438	.06316	.06195	.06077	.05959	.05844	.05730	.05618	.05508
2.0	.05399	.05292	.05186	.05082	.04980	.04879	.04780	.04682	.04586	.04491
2.1	.04398	.04307	.04217	.04128	.04041	.03955	.03871	.03788	.03706	.03626
2.2	.03547	.03470	.03394	.03319	.03246	.03174	.03103	.03034	.02965	.02898
2.3	.02833	.02768	.02705	.02643	.02582	.02522	.02463	.02406	.02349	.02294
2.4	.02239	.02186	.02134	.02083	.02033	.01984	.01936	.01888	.01842	.01797
2.5	.01753	.01709	.01667	.01625	.01585	.01545	.01506	.01468	.01431	.01394
2.6	.01358	.01323	.01289	.01256	.01223	.01191	.01160	.01130	.01100	.01071
2.7	.01042	.01014	.00987	.00961	.00935	.00909	.00885	.00861	.00837	.00814
2.8	.00792	.00770	.00748	.00727	.00707	.00687	.00668	.00649	.00631	.00613
2.9	.00595	.00578	.00562	.00545	.00530	.00514	.00499	.00485	.00470	.00457
3.0	.00443	.00430	.00417	.00405	.00393	.00381	.00370	.00358	.00348	.00337
3.1	.00327	.00317	.00307	.00298	.00288	.00279	.00271	.00262	.00254	.00246
3.2	.00238	.00231	.00224	.00216	.00210	.00203	.00196	.00190	.00184	.00178
3.3	.00172	.00167	.00161	.00156	.00151	.00146	.00141	.00136	.00132	.00127
3.4	.00123	.00119	.00115	.00111	.00107	.00104	.00100	.00097	.00094	.00090
3.5	.00087	.00084	.00081	.00079	.00076	.00073	.00071	.00068	.00066	.00063
3.6	.00061	.00059	.00057	.00055	.00053	.00051	.00049	.00047	.00046	.00044
3.7	.00042	.00041	.00039	.00038	.00037	.00035	.00034	.00033	.00031	.00030
3.8	.00029	.00028	.00027	.00026	.00025	.00024	.00023	.00022	.00021	.00021
3.9	.00020	.00019	.00018	.00018	.00017	.00016	.00016	.00015	.00014	.00014
4.0	.00013	.00013	.00012	.00012	.00011	.00011	.00011	.00010	.00010	.00009

Table 13. χ^2 (Chi-squared)-distribution

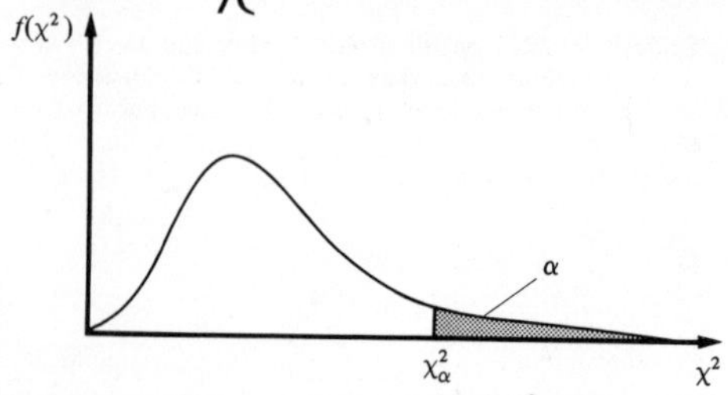

Values of χ^2_α giving area (α) in the right-hand tail for different number of degrees of freedom (ν). **Example:** For $\nu = 15$ area beyond $\chi^2_{0.95} = 7.261$ is 0.95 and beyond $\chi^2_{0.10} = 22.307$ is 0.10.

ν \ α	0.995	0.990	0.975	0.950	0.900	0.750	0.500	0.250	0.100	0.050	0.025	0.010	0.005	0.001
1	$0.0^{4}3927$*	$0.0^{3}1571$*	$0.0^{3}9821$*	$0.0^{2}3932$*	0.01579	0.1015	0.4549	1.323	2.706	3.841	5.024	6.635	7.879	10.828
2	0.01003	0.02010	0.05064	0.1026	0.2107	0.5754	1.386	2.773	4.605	5.991	7.378	9.210	10.597	13.816
3	0.07172	0.1148	0.2158	0.3518	0.5844	1.213	2.366	4.108	6.251	7.815	9.348	11.345	12.838	16.266
4	0.2070	0.2971	0.4844	0.7107	1.064	1.923	3.357	5.385	7.779	9.488	11.143	13.277	14.860	18.467
5	0.4117	0.5543	0.8312	1.145	1.610	2.675	4.351	6.626	9.236	11.070	12.833	15.086	16.750	20.515
6	0.6757	0.8721	1.237	1.635	2.204	3.455	5.348	7.841	10.645	12.592	14.449	16.812	18.548	22.458
7	0.9893	1.239	1.690	2.167	2.833	4.255	6.346	9.037	12.017	14.067	16.013	18.475	20.278	24.322
8	1.344	1.646	2.180	2.733	3.490	5.071	7.344	10.219	13.362	15.507	17.535	20.090	21.955	26.125
9	1.735	2.088	2.700	3.325	4.168	5.899	8.343	11.389	14.684	16.919	19.023	21.666	23.589	27.877
10	2.156	2.558	3.247	3.940	4.865	6.737	9.342	12.549	15.987	18.307	20.483	23.209	25.188	29.588
11	2.603	3.053	3.816	4.575	5.578	7.584	10.341	13.701	17.275	19.675	21.920	24.725	26.757	31.264
12	3.074	3.571	4.404	5.226	6.304	8.438	11.340	14.845	18.549	21.026	23.337	26.217	28.300	32.909
13	3.565	4.107	5.009	5.892	7.041	9.299	12.340	15.984	19.812	22.362	24.736	27.688	29.819	34.528
14	4.075	4.660	5.629	6.571	7.790	10.165	13.339	17.117	21.064	23.685	26.119	29.141	31.319	36.123
15	4.601	5.229	6.262	7.261	8.547	11.036	14.339	18.245	22.307	24.996	27.488	30.578	32.801	37.697
16	5.142	5.812	6.908	7.962	9.312	11.912	15.338	19.369	23.542	26.296	28.845	32.000	34.267	39.252
17	5.697	6.408	7.564	8.672	10.085	12.792	16.338	20.489	24.769	27.587	30.191	33.409	35.718	40.790
18	6.265	7.015	8.231	9.390	10.865	13.675	17.338	21.605	25.989	28.869	31.526	34.805	37.156	42.312
19	6.844	7.633	8.907	10.117	11.651	14.562	18.338	22.718	27.204	30.143	32.852	36.191	38.582	43.820
20	7.434	8.260	9.591	10.851	12.443	15.452	19.337	23.828	28.412	31.410	34.170	37.566	39.997	45.315
21	8.034	8.897	10.283	11.591	13.240	16.344	20.337	24.935	29.615	32.670	35.479	38.932	41.401	46.797
22	8.643	9.542	10.982	12.338	14.041	17.240	21.337	26.039	30.813	33.924	36.781	40.289	42.796	48.268
23	9.260	10.196	11.688	13.090	14.848	18.137	22.337	27.141	32.007	35.172	38.076	41.638	44.181	49.728
24	9.886	10.856	12.401	13.848	15.659	19.037	23.337	28.241	33.196	36.415	39.364	42.080	45.558	51.179
25	10.520	11.524	13.120	14.611	16.473	19.939	24.337	29.339	34.382	37.652	40.646	44.314	46.928	52.620
26	11.160	12.198	13.844	15.379	17.292	20.843	25.336	30.434	35.563	38.885	41.923	45.642	48.290	54.052
27	11.808	12.879	14.573	16.151	18.114	21.749	26.336	31.528	36.741	40.113	43.194	46.963	49.645	55.476
28	12.461	13.565	15.308	16.928	18.939	22.657	27.336	32.620	37.916	41.337	44.461	48.278	50.993	56.892
29	13.121	14.256	16.047	17.708	19.768	23.567	28.336	33.711	39.087	42.557	45.722	49.588	52.336	58.302
30	13.787	14.954	16.791	18.493	20.599	24.478	29.336	34.800	40.256	43.773	46.979	50.892	53.672	59.703
40	20.707	22.164	24.433	26.509	29.050	33.660	39.335	45.616	51.805	55.758	59.342	63.691	66.766	73.402
50	27.991	29.707	32.357	34.764	37.689	42.942	49.335	56.334	63.167	67.505	71.420	76.154	79.490	86.661
60	35.535	37.485	40.482	43.188	46.459	52.294	59.335	66.981	74.397	79.082	83.298	88.379	91.952	99.607
70	43.275	45.442	48.758	51.739	55.329	61.698	69.334	77.577	85.527	90.531	95.023	100.425	104.215	112.317
80	51.172	53.540	57.153	60.391	64.278	71.144	79.334	88.130	96.578	101.879	106.629	112.329	116.321	124.839
90	59.196	61.754	65.647	69.126	73.291	80.625	89.334	98.650	107.565	113.145	118.136	124.116	128.299	137.208
100	67.328	70.065	74.222	77.929	82.358	90.133	99.334	109.141	118.498	124.342	129.561	135.807	140.169	149.449

z_α	−2.5758	−2.3263	−1.9600	−1.6449	−1.2816	−0.6745	0.0000	0.6745	1.2816	1.6449	1.9600	2.3263	2.5758	3.0902

* e.g. $0.0^{4}3927 = 0.00003927$

Interpolation: For $\nu > 100$ $\chi^2_\alpha = \frac{1}{2}(z_\alpha + \sqrt{2\nu - 1})^2$ where z_α is the standardised Normal variable shown in the bottom line of the table.

Reproduced by permission from the *Biometrika Tables for Statisticians*, Vol. 1, Table 8.

Table 14. *t*-distribution

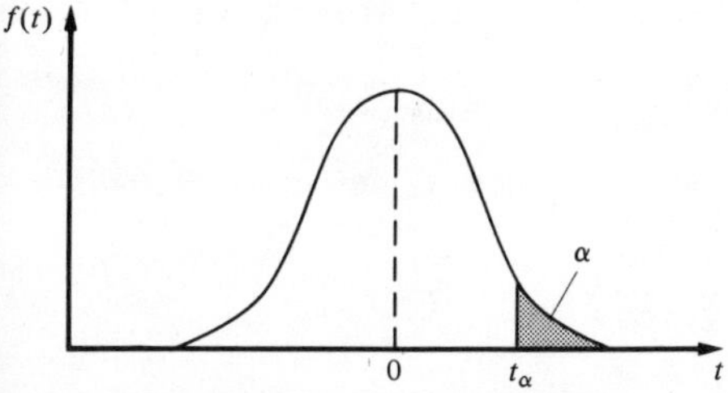

Critical points (t_α) for different probability levels (α) and different number of degrees of freedom (ν). **Example**: For $\nu = 19$, $P(t > 2.0930) = 0.025$ and $P(|t| > 2.0930) = 0.05$.

ν \ α	0.4	0.25	0.15	0.1	0.05	0.025	0.01	0.005	0.001	0.0005
1	0.3249	1.0000	1.9626	3.0777	6.3138	12.7062	31.8205	63.6567	318.3087	636.6189
2	0.2887	0.8165	1.3862	1.8856	2.9200	4.3027	6.9646	9.9248	22.3271	31.5991
3	0.2767	0.7649	1.2498	1.6377	2.3534	3.1824	4.5407	5.8409	10.2145	12.9240
4	0.2707	0.7407	1.1896	1.5332	2.1318	2.7764	3.7469	4.6041	7.1732	8.6103
5	0.2672	0.7267	1.1558	1.4759	2.0150	2.5706	3.3649	4.0321	5.8934	6.8688
6	0.2648	0.7176	1.1342	1.4398	1.9432	2.4469	3.1427	3.7074	5.2076	5.9588
7	0.2632	0.7111	1.1192	1.4149	1.8946	2.3646	2.9980	3.4995	4.7853	5.4079
8	0.2619	0.7064	1.1081	1.3968	1.8595	2.3060	2.8965	3.3554	4.5008	5.0413
9	0.2610	0.7027	1.0997	1.3830	1.8331	2.2622	2.8214	3.2498	4.2968	4.7809
10	0.2602	0.6998	1.0931	1.3722	1.8125	2.2281	2.7638	3.1693	4.1437	4.5869
11	0.2596	0.6974	1.0877	1.3634	1.7959	2.2010	2.7181	3.1058	4.0247	4.4370
12	0.2590	0.6955	1.0832	1.3562	1.7823	2.1788	2.6810	3.0545	3.9296	4.3178
13	0.2586	0.6938	1.0795	1.3502	1.7709	2.1604	2.6503	3.0123	3.8520	4.2208
14	0.2582	0.6924	1.0763	1.3450	1.7613	2.1448	2.6245	2.9768	3.7874	4.1405
15	0.2579	0.6912	1.0735	1.3406	1.7531	2.1314	2.6025	2.9467	3.7328	4.0728
16	0.2576	0.6901	1.0711	1.3368	1.7459	2.1199	2.5835	2.9208	3.6862	4.0150
17	0.2573	0.6892	1.0690	1.3334	1.7396	2.1098	2.5669	2.8982	3.6458	3.9651
18	0.2571	0.6884	1.0672	1.3304	1.7341	2.1009	2.5524	2.8784	3.6105	3.9216
19	0.2569	0.6876	1.0655	1.3277	1.7291	2.0930	2.5395	2.8609	3.5794	3.8834
20	0.2567	0.6870	1.0640	1.3253	1.7247	2.0860	2.5280	2.8453	3.5518	3.8495
21	0.2566	0.6864	1.0627	1.3232	1.7207	2.0796	2.5176	2.8314	3.5272	3.8193
22	0.2564	0.6858	1.0614	1.3212	1.7171	2.0739	2.5083	2.8188	3.5050	3.7921
23	0.2563	0.6853	1.0603	1.3195	1.7139	2.0687	2.4999	2.8073	3.4850	3.7676
24	0.2562	0.6848	1.0593	1.3178	1.7109	2.0639	2.4922	2.7969	3.4668	3.7454
25	0.2561	0.6844	1.0584	1.3163	1.7081	2.0595	2.4851	2.7874	3.4502	3.7251
26	0.2560	0.6840	1.0575	1.3150	1.7056	2.0555	2.4786	2.7787	3.4350	3.7066
27	0.2559	0.6837	1.0567	1.3137	1.7033	2.0518	2.4727	2.7707	3.4210	3.6896
28	0.2558	0.6834	1.0560	1.3125	1.7011	2.0484	2.4671	2.7633	3.4082	3.6739
29	0.2557	0.6830	1.0553	1.3114	1.6991	2.0452	2.4620	2.7564	3.3962	3.6594
30	0.2556	0.6828	1.0547	1.3104	1.6973	2.0423	2.4573	2.7500	3.3852	3.6460
35	0.2553	0.6816	1.0520	1.3062	1.6896	2.0301	2.4377	2.7238	3.3400	3.5911
40	0.2550	0.6807	1.0500	1.3031	1.6839	2.0211	2.4233	2.7045	3.3069	3.5510
45	0.2549	0.6800	1.0485	1.3006	1.6794	2.0141	2.4121	2.6896	3.2815	3.5203
50	0.2547	0.6794	1.0473	1.2987	1.6759	2.0086	2.4033	2.6778	3.2614	3.4960
60	0.2545	0.6786	1.0455	1.2958	1.6706	2.0003	2.3901	2.6603	3.2317	3.4602
70	0.2543	0.6780	1.0442	1.2938	1.6669	1.9944	2.3808	2.6479	3.2108	3.4350
80	0.2542	0.6776	1.0432	1.2922	1.6641	1.9901	2.3739	2.6387	3.1953	3.4163
90	0.2541	0.6772	1.0424	1.2910	1.6620	1.9867	2.3685	2.6316	3.1833	3.4019
100	0.2540	0.6770	1.0418	1.2901	1.6602	1.9840	2.3642	2.6259	3.1737	3.3905
120	0.2539	0.6765	1.0409	1.2886	1.6577	1.9799	2.3578	2.6174	3.1595	3.3735
150	0.2538	0.6761	1.0400	1.2872	1.6551	1.9759	2.3515	2.6090	3.1455	3.3566
200	0.2537	0.6757	1.0391	1.2858	1.6525	1.9719	2.3451	2.6006	3.1315	3.3398
300	0.2536	0.6753	1.0382	1.2844	1.6499	1.9679	2.3388	2.5923	3.1176	3.3233
∞	0.2533	0.6745	1.0364	1.2816	1.6449	1.9600	2.3263	2.5758	3.0902	3.2905

Table 15. *F*-distribution

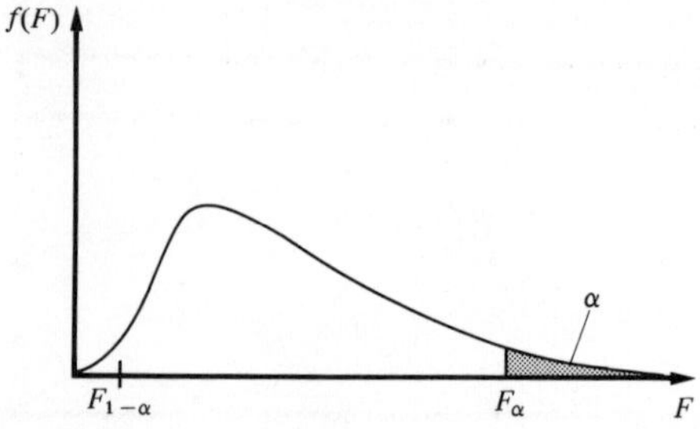

Values of F_α ($\alpha = 0.1$, 0.05 and 0.01) for different combinations of degrees of freedom in the numerator, ν_1, and denominator, ν_2. Example: When $\nu_1 = 10$ and $\nu_2 = 20$ area (α) to the right of $F_{0.05} = 2.348$ is 0.05. To find $F_{1-\alpha}$, leaving an area α in the left-hand tail, use the relation: $F_{1-\alpha}(\nu_1, \nu_2) = 1/F_\alpha(\nu_2, \nu_1)$. Example: $F_{0.95}(20, 10) = 1/F_{0.05}(10, 20) = 1/2.348 = 0.4259$.

ν_2	α \ ν_1	1	2	3	4	5	6	7	8	9	10	12	15	20	25	30	50	100	∞
1	0.10	39.86	49.50	53.59	55.83	57.24	58.20	58.91	59.44	59.86	60.19	60.71	61.22	61.74	62.05	62.26	62.69	63.01	63.33
	0.05	161.4	199.5	215.7	224.6	230.2	234.0	236.8	238.9	240.5	241.9	243.9	245.9	248.0	249.3	250.1	251.8	253.0	254.3
	0.01	4052	4999	5403	5625	5764	5859	5928	5981	6022	6056	6106	6157	6209	6240	6261	6303	6334	6366
2	0.10	8.526	9.000	9.162	9.243	9.293	9.326	9.349	9.367	9.381	9.392	9.408	9.425	9.441	9.451	9.458	9.471	9.481	9.491
	0.05	18.51	19.00	19.16	19.25	19.30	19.33	19.35	19.37	19.38	19.40	19.41	19.43	19.45	19.46	19.46	19.48	19.49	19.50
	0.01	98.50	99.00	99.17	99.25	99.30	99.33	99.36	99.37	99.39	99.40	99.42	99.43	99.45	99.46	99.47	99.48	99.49	99.50
3	0.10	5.538	5.462	5.391	5.343	5.309	5.285	5.266	5.252	5.240	5.230	5.216	5.200	5.184	5.175	5.168	5.155	5.144	5.134
	0.05	10.13	9.552	9.277	9.117	9.013	8.941	8.887	8.845	8.812	8.786	8.745	8.703	8.660	8.634	8.617	8.581	8.554	8.526
	0.01	34.12	30.82	29.46	28.71	28.24	27.91	27.67	27.49	27.35	27.23	27.05	26.87	26.69	26.58	26.50	26.35	26.24	26.13
4	0.10	4.545	4.325	4.191	4.107	4.051	4.010	3.979	3.955	3.936	3.920	3.896	3.870	3.844	3.828	3.817	3.795	3.778	3.761
	0.05	7.709	6.944	6.591	6.388	6.256	6.163	6.094	6.041	5.999	5.964	5.912	5.858	5.803	5.769	5.746	5.699	5.664	5.628
	0.01	21.20	18.00	16.69	15.98	15.52	15.21	14.98	14.80	14.66	14.55	14.37	14.20	14.02	13.91	13.84	13.69	13.58	13.46
5	0.10	4.060	3.780	3.619	3.520	3.453	3.405	3.368	3.339	3.316	3.297	3.268	3.238	3.207	3.187	3.174	3.147	3.126	3.105
	0.05	6.608	5.786	5.409	5.192	5.050	4.950	4.876	4.818	4.772	4.735	4.678	4.619	4.558	4.521	4.496	4.444	4.405	4.365
	0.01	16.26	13.27	12.06	11.39	10.97	10.67	10.46	10.29	10.16	10.05	9.888	9.722	9.553	9.449	9.379	9.238	9.130	9.020
6	0.10	3.776	3.463	3.289	3.181	3.108	3.055	3.014	2.983	2.958	2.937	2.905	2.871	2.836	2.815	2.800	2.770	2.746	2.722
	0.05	5.987	5.143	4.757	4.534	4.387	4.284	4.207	4.147	4.099	4.060	4.000	3.938	3.874	3.835	3.808	3.754	3.712	3.669
	0.01	13.75	10.92	9.780	9.148	8.746	8.466	8.260	8.102	7.976	7.874	7.718	7.559	7.396	7.296	7.229	7.091	6.987	6.880
7	0.10	3.589	3.257	3.074	2.961	2.883	2.827	2.785	2.752	2.725	2.703	2.668	2.632	2.595	2.571	2.555	2.523	2.497	2.471
	0.05	5.591	4.737	4.347	4.120	3.972	3.866	3.787	3.726	3.677	3.637	3.575	3.511	3.445	3.404	3.376	3.319	3.275	3.230
	0.01	12.25	9.547	8.451	7.847	7.460	7.191	6.993	6.840	6.719	6.620	6.469	6.314	6.155	6.058	5.992	5.858	5.755	5.650
8	0.10	3.458	3.113	2.924	2.806	2.726	2.668	2.624	2.589	2.561	2.538	2.502	2.464	2.425	2.400	2.383	2.348	2.321	2.293
	0.05	5.318	4.459	4.066	3.838	3.687	3.581	3.500	3.438	3.388	3.347	3.284	3.218	3.150	3.108	3.079	3.020	2.975	2.928
	0.01	11.26	8.649	7.591	7.006	6.632	6.371	6.178	6.029	5.911	5.814	5.667	5.515	5.359	5.263	6.198	5.065	4.963	4.859
9	0.10	3.360	3.006	2.813	2.693	2.611	2.551	2.505	2.469	2.440	2.416	2.379	2.340	2.298	2.272	2.255	2.218	2.189	2.159
	0.05	5.117	4.256	3.863	3.633	3.482	3.374	3.293	3.230	3.179	3.137	3.073	3.006	2.936	2.893	2.864	2.803	2.756	2.707
	0.01	10.56	8.022	6.992	6.422	6.057	5.802	5.613	5.467	5.351	5.257	5.111	4.962	4.808	4.713	4.649	4.517	4.415	4.311
10	0.10	3.285	2.924	2.728	2.605	2.522	2.461	2.414	2.377	2.347	2.323	2.284	2.244	2.201	2.174	2.155	2.117	2.087	2.055
	0.05	4.965	4.103	3.708	3.478	3.326	3.217	3.135	3.072	3.020	2.978	2.913	2.845	2.774	2.730	2.700	2.637	2.588	2.538
	0.01	10.04	7.559	6.552	5.994	5.636	5.386	5.200	5.057	4.942	4.849	4.706	4.558	4.405	4.311	4.247	4.115	4.014	3.909
11	0.10	3.225	2.860	2.660	2.536	2.451	2.389	2.342	2.304	2.274	2.248	2.209	2.167	2.123	2.095	2.076	2.036	2.005	1.972
	0.05	4.844	3.982	3.587	3.357	3.204	3.095	3.012	2.948	2.896	2.854	2.788	2.719	2.646	2.601	2.570	2.507	2.457	2.404
	0.01	9.646	7.206	6.217	5.668	5.316	5.069	4.886	4.744	4.632	4.539	4.397	4.251	4.099	4.005	3.941	3.810	3.708	3.602
12	0.10	3.177	2.807	2.606	2.480	2.394	2.331	2.283	2.245	2.214	2.188	2.147	2.105	2.060	2.031	2.011	1.970	1.938	1.904
	0.05	4.747	3.885	3.490	3.259	3.106	2.996	2.913	2.849	2.796	2.753	2.687	2.617	2.544	2.498	2.466	2.401	2.350	2.296
	0.01	9.330	6.927	5.953	5.412	5.064	4.821	4.640	4.499	4.388	4.296	4.155	4.010	3.858	3.765	3.701	3.569	3.467	3.361
13	0.10	3.136	2.763	2.560	2.434	2.347	2.283	2.234	2.195	2.164	2.138	2.097	2.053	2.007	1.978	1.958	1.915	1.882	1.846
	0.05	4.667	3.806	3.411	3.179	3.025	2.915	2.832	2.767	2.714	2.671	2.604	2.533	2.459	2.412	2.380	2.314	2.261	2.206
	0.01	9.074	6.701	5.739	5.205	4.862	4.620	4.441	4.302	4.191	4.100	3.960	3.815	3.665	3.571	3.507	3.375	3.272	3.165
14	0.10	3.102	2.726	2.522	2.395	2.307	2.243	2.193	2.154	2.122	2.095	2.054	2.010	1.962	1.933	1.912	1.869	1.834	1.797
	0.05	4.600	3.739	3.344	3.112	2.958	2.848	2.764	2.699	2.646	2.602	2.534	2.463	2.388	2.341	2.308	2.241	2.187	2.131
	0.01	8.862	6.515	5.564	5.035	4.695	4.456	4.278	4.140	4.030	3.939	3.800	3.656	3.505	3.412	3.348	3.215	3.112	3.004
15	0.10	3.073	2.695	2.490	2.361	2.273	2.208	2.158	2.119	2.086	2.059	2.017	1.972	1.924	1.894	1.873	1.828	1.793	1.755
	0.05	4.543	3.682	3.287	3.056	2.901	2.790	2.707	2.641	2.588	2.544	2.475	2.403	2.328	2.280	2.247	2.178	2.123	2.066
	0.01	8.683	6.359	5.417	4.893	4.556	4.318	4.142	4.004	3.895	3.805	3.666	3.522	3.372	3.278	3.214	3.081	2.977	2.868
16	0.10	3.048	2.668	2.462	2.333	2.244	2.178	2.128	2.088	2.055	2.028	1.985	1.940	1.891	1.860	1.839	1.793	1.757	1.718
	0.05	4.494	3.634	3.239	3.007	2.852	2.741	2.657	2.591	2.538	2.494	2.425	2.352	2.276	2.227	2.194	2.124	2.068	2.010
	0.01	8.531	6.226	5.292	4.773	4.437	4.202	4.026	3.890	3.780	3.691	3.553	3.409	3.259	3.165	3.101	2.967	2.863	2.753
17	0.10	3.026	2.645	2.437	2.308	2.218	2.152	2.102	2.061	2.028	2.001	1.958	1.912	1.862	1.831	1.809	1.763	1.726	1.686
	0.05	4.451	3.592	3.197	2.965	2.810	2.699	2.614	2.548	2.494	2.450	2.381	2.308	2.230	2.181	2.148	2.077	2.020	1.960
	0.01	8.400	6.112	5.185	4.669	4.336	4.102	3.927	3.791	3.682	3.593	3.455	3.312	3.162	3.068	3.003	2.869	2.764	2.653
18	0.10	3.007	2.624	2.416	2.286	2.196	2.130	2.079	2.038	2.005	1.997	1.933	1.887	1.837	1.805	1.783	1.736	1.698	1.657
	0.05	4.414	3.555	3.160	2.928	2.773	2.661	2.577	2.510	2.456	2.412	2.342	2.269	2.191	2.141	2.107	2.035	1.978	1.917
	0.01	8.285	6.013	5.092	4.579	4.248	4.015	3.841	3.705	3.597	3.508	3.371	3.227	3.077	2.983	2.919	2.784	2.678	2.566
19	0.10	2.990	2.606	2.397	2.266	2.176	2.109	2.058	2.017	1.984	1.956	1.912	1.865	1.814	1.782	1.759	1.711	1.673	1.631
	0.05	4.381	3.522	3.127	2.895	2.740	2.628	2.544	2.477	2.423	2.378	2.308	2.234	2.155	2.106	2.071	1.999	1.940	1.878
	0.01	8.185	5.926	5.010	4.500	4.171	3.939	3.765	3.631	3.523	3.434	3.297	3.153	3.003	2.909	2.844	2.709	2.602	2.489
20	0.10	2.975	2.589	2.380	2.249	2.158	2.091	2.040	1.999	1.965	1.937	1.892	1.845	1.794	1.761	1.738	1.690	1.650	1.607
	0.05	4.351	3.493	3.098	2.866	2.711	2.599	2.514	2.447	2.393	2.348	2.278	2.203	2.124	2.074	2.039	1.966	1.907	1.843
	0.01	8.096	5.849	4.938	4.431	4.103	3.871	3.699	3.564	3.457	3.368	3.231	3.088	2.938	2.843	2.778	2.643	2.535	2.421

Table 15 – *continued*

ν_2	α \ ν_1	1	2	3	4	5	6	7	8	9	10	12	15	20	25	30	50	100	∞
21	0.10	2.961	2.575	2.365	2.233	2.142	2.075	2.023	1.982	1.948	1.920	1.875	1.827	1.776	1.742	1.719	1.670	1.630	1.586
	0.05	4.325	3.467	3.072	2.840	2.685	2.573	2.488	2.420	2.366	2.321	2.250	2.176	2.096	2.045	2.010	1.936	1.876	1.812
	0.01	8.017	5.780	4.874	4.369	4.042	3.812	3.640	3.506	3.398	3.310	3.173	3.030	2.880	2.785	2.720	2.584	2.475	2.360
22	0.10	2.949	2.561	2.351	2.219	2.128	2.060	2.008	1.967	1.933	1.904	1.859	1.811	1.759	1.726	1.702	1.652	1.611	1.567
	0.05	4.301	3.443	3.049	2.817	2.661	2.549	2.464	2.397	2.342	2.297	2.226	2.151	2.071	2.020	1.984	1.909	1.849	1.783
	0.01	7.945	5.719	4.817	4.313	3.988	3.758	3.587	3.453	3.346	3.258	3.121	2.978	2.827	2.733	2.667	2.531	2.422	2.305
23	0.10	2.937	2.549	2.339	2.207	2.115	2.047	1.995	1.953	1.919	1.890	1.845	1.796	1.744	1.710	1.686	1.636	1.594	1.549
	0.05	4.279	3.422	3.028	2.796	2.640	2.528	2.442	2.375	2.320	2.275	2.204	2.128	2.048	1.996	1.961	1.885	1.823	1.757
	0.01	7.881	5.664	4.765	4.264	3.939	3.710	3.539	3.406	3.299	3.211	3.074	2.931	2.781	2.686	2.620	2.483	2.373	2.256
24	0.10	2.927	2.538	2.327	2.195	2.103	2.035	1.983	1.941	1.906	1.877	1.832	1.783	1.730	1.696	1.672	1.621	1.579	1.533
	0.05	4.260	3.403	3.009	2.776	2.621	2.508	2.423	2.355	2.300	2.255	2.183	2.108	2.027	1.975	1.939	1.863	1.800	1.733
	0.01	7.823	5.614	4.718	4.218	3.895	3.667	3.496	3.363	3.256	3.168	3.032	2.889	2.738	2.643	2.577	2.440	2.329	2.211
25	0.10	2.918	2.528	2.317	2.184	2.092	2.024	1.971	1.929	1.895	1.866	1.820	1.771	1.718	1.683	1.659	1.607	1.565	1.518
	0.05	4.242	3.385	2.991	2.759	2.603	2.490	2.405	2.337	2.282	2.236	2.165	2.089	2.007	1.955	1.919	1.842	1.779	1.711
	0.01	7.770	5.568	4.675	4.177	3.855	3.627	3.457	3.324	3.217	3.129	2.993	2.850	2.699	2.604	2.538	2.400	2.289	2.169
26	0.10	2.909	2.519	2.307	2.174	2.082	2.014	1.961	1.919	1.884	1.855	1.809	1.760	1.706	1.671	1.647	1.594	1.551	1.504
	0.05	4.225	3.369	2.975	2.743	2.587	2.474	2.388	2.321	2.265	2.220	2.148	2.072	1.990	1.938	1.901	1.823	1.760	1.691
	0.01	7.721	5.526	4.637	4.140	3.818	3.591	3.421	3.288	3.182	3.094	2.958	2.815	2.664	2.569	2.503	2.364	2.252	2.131
27	0.10	2.901	2.511	2.299	2.165	2.073	2.005	1.952	1.909	1.874	1.845	1.799	1.749	1.695	1.660	1.636	1.583	1.539	1.491
	0.05	4.210	3.354	2.960	2.728	2.572	2.459	2.373	2.305	2.250	2.204	2.132	2.056	1.974	1.921	1.884	1.806	1.742	1.672
	0.01	7.677	5.488	4.601	4.106	3.785	3.558	3.388	3.256	3.149	3.062	2.926	2.783	2.632	2.536	2.470	2.330	2.218	2.097
28	0.10	2.894	2.503	2.291	2.157	2.064	1.996	1.943	1.900	1.865	1.836	1.790	1.740	1.685	1.650	1.625	1.572	1.528	1.478
	0.05	4.196	3.340	2.947	2.714	2.558	2.445	2.359	2.291	2.236	2.190	2.118	2.041	2.959	1.906	1.869	1.790	1.725	1.654
	0.01	7.636	5.453	4.568	4.074	3.754	3.528	3.358	3.226	3.120	3.032	2.896	2.753	2.602	2.506	2.440	2.300	2.187	2.064
29	0.10	2.887	2.495	2.283	2.149	2.057	1.988	1.935	1.892	1.857	1.827	1.781	1.731	1.676	1.640	1.616	1.562	1.517	1.467
	0.05	4.183	3.328	2.934	2.701	2.545	2.432	2.346	2.278	2.223	2.177	2.104	2.027	1.945	1.891	1.854	1.775	1.710	1.638
	0.01	7.598	5.420	4.538	4.045	3.725	3.499	3.330	3.198	3.092	3.005	2.868	2.726	2.574	2.478	2.412	2.271	2.158	2.034
30	0.10	2.881	2.489	2.276	2.142	2.049	1.980	1.927	1.884	1.849	1.819	1.773	1.722	1.667	1.632	1.606	1.552	1.507	1.456
	0.05	4.171	3.316	2.922	2.690	2.534	2.421	2.334	2.266	2.211	2.165	2.092	2.015	1.932	1.878	1.841	1.761	1.695	1.622
	0.01	7.562	5.390	4.510	4.018	3.699	3.473	3.304	3.173	3.067	2.979	2.843	2.700	2.549	2.453	2.386	2.245	2.131	2.006
35	0.10	2.855	2.461	2.247	2.113	2.019	1.950	1.896	1.852	1.817	1.787	1.739	1.688	1.632	1.595	1.569	1.513	1.465	1.411
	0.05	4.121	3.267	2.874	2.641	2.485	2.372	2.285	2.217	2.161	2.114	2.041	1.963	1.878	1.824	1.786	1.703	1.635	1.558
	0.01	7.419	5.268	4.396	3.908	3.592	3.368	3.200	3.069	2.963	2.876	2.740	2.597	2.445	2.348	2.281	2.137	2.020	1.891
40	0.10	2.835	2.440	2.226	2.091	1.997	1.927	1.873	1.829	1.793	1.763	1.715	1.662	1.605	1.568	1.541	1.483	1.434	1.377
	0.05	4.085	3.232	2.839	2.606	2.449	2.336	2.249	2.180	2.124	2.077	2.003	1.924	1.839	1.783	1.744	1.660	1.589	1.509
	0.01	7.314	5.170	4.313	3.828	3.514	3.291	3.124	2.993	2.888	2.801	2.665	2.522	2.369	2.271	2.203	2.058	1.938	1.805
45	0.10	2.820	2.425	2.210	2.074	1.980	1.909	1.855	1.811	1.774	1.744	1.695	1.643	1.585	1.546	1.519	1.460	1.409	1.349
	0.05	4.057	3.204	2.812	2.579	2.422	2.308	2.221	2.152	2.096	2.049	1.974	1.895	1.808	1.752	1.713	1.626	1.554	1.470
	0.01	7.234	5.110	4.249	3.767	3.454	3.232	3.066	2.935	2.830	2.743	2.608	2.464	2.311	2.213	2.144	1.997	1.875	1.737
50	0.10	2.809	2.412	2.197	2.061	1.966	1.895	1.840	1.796	1.760	1.729	1.680	1.627	1.568	1.529	1.502	1.441	1.388	1.327
	0.05	4.034	3.183	2.790	2.557	2.400	2.286	2.199	2.130	2.073	2.026	1.952	1.871	1.784	1.727	1.687	1.599	1.525	1.438
	0.01	7.171	5.057	4.199	3.720	3.408	3.186	3.020	2.890	2.785	2.698	2.562	2.419	2.265	2.167	2.098	1.949	1.825	1.683
60	0.10	2.791	2.393	2.177	2.041	1.946	1.875	1.819	1.775	1.738	1.707	1.657	1.603	1.543	1.504	1.476	1.413	1.358	1.291
	0.05	4.001	3.150	2.758	2.525	2.368	2.254	2.167	2.097	2.040	1.993	1.917	1.836	1.748	1.690	1.649	1.559	1.481	1.389
	0.01	7.077	4.977	4.126	3.649	3.339	3.119	2.953	2.823	2.718	2.632	2.496	2.352	2.198	2.098	2.028	1.877	1.749	1.601
70	0.10	2.779	2.380	2.164	2.027	1.931	1.860	1.804	1.760	1.723	1.691	1.641	1.587	1.526	1.486	1.457	1.392	1.335	1.265
	0.05	3.978	3.128	2.736	2.503	2.346	2.231	2.143	2.074	2.017	1.969	1.893	1.812	1.722	1.664	1.622	1.530	1.450	1.353
	0.01	7.011	4.922	4.074	3.600	3.291	3.071	2.906	2.777	2.672	2.585	2.450	2.306	2.150	2.050	1.980	1.826	1.695	1.540
80	0.10	2.769	2.370	2.154	2.016	1.921	1.849	1.793	1.748	1.711	1.680	1.629	1.574	1.513	1.472	1.443	1.377	1.318	1.245
	0.05	3.960	3.111	2.719	2.486	2.329	2.214	2.126	2.056	1.999	1.951	1.875	1.793	1.703	1.644	1.602	1.508	1.426	1.325
	0.01	6.963	4.881	4.036	3.563	3.255	3.036	2.871	2.742	2.637	2.551	2.415	2.271	2.115	2.015	1.944	1.788	1.655	1.494
90	0.10	2.762	2.363	2.146	2.008	1.912	1.841	1.785	1.739	1.702	1.670	1.620	1.564	1.503	1.461	1.432	1.365	1.304	1.228
	0.05	3.947	3.098	2.706	2.473	2.316	2.201	2.113	2.043	1.986	1.938	1.861	1.779	1.688	1.629	1.586	1.491	1.407	1.302
	0.01	6.925	4.849	4.007	3.535	3.228	3.009	2.845	2.715	2.611	2.524	2.389	2.244	2.088	1.987	1.916	1.759	1.623	1.457
100	0.10	2.756	2.356	2.139	2.002	1.906	1.834	1.778	1.732	1.695	1.663	1.612	1.557	1.494	1.453	1.423	1.355	1.293	1.214
	0.05	3.936	3.087	2.696	2.463	2.305	2.191	2.103	2.032	1.975	1.927	1.850	1.768	1.676	1.616	1.573	1.477	1.392	1.283
	0.01	6.895	4.824	3.984	3.513	3.206	2.988	2.823	2.694	2.590	2.503	2.368	2.223	2.067	1.965	1.893	1.735	1.598	1.427
120	0.10	2.748	2.347	2.130	1.992	1.896	1.824	1.767	1.722	1.684	1.652	1.601	1.545	1.482	1.440	1.409	1.340	1.277	1.193
	0.05	3.920	3.072	2.680	2.447	2.290	2.175	2.087	2.016	1.959	1.910	1.834	1.750	1.659	1.598	1.554	1.457	1.369	1.254
	0.01	6.851	4.787	3.949	3.480	3.174	2.956	2.792	2.663	2.559	2.472	2.336	2.192	2.035	1.932	1.860	1.700	1.559	1.381
150	0.10	2.739	2.338	2.121	1.983	1.886	1.814	1.757	1.712	1.674	1.642	1.590	1.533	1.470	1.427	1.396	1.325	1.259	1.169
	0.05	3.904	3.056	2.665	2.432	2.274	2.160	2.071	2.001	1.943	1.894	1.817	1.734	1.641	1.580	1.535	1.436	1.345	1.223
	0.01	6.807	4.749	3.915	3.447	3.142	2.924	2.761	2.632	2.528	2.441	2.305	2.160	2.003	1.900	1.827	1.665	1.520	1.331
200	0.10	2.731	2.329	2.111	1.973	1.876	1.804	1.747	1.701	1.663	1.631	1.579	1.522	1.458	1.414	1.383	1.310	1.242	1.144
	0.05	3.888	3.041	2.650	2.417	2.259	2.144	2.056	1.985	1.927	1.878	1.801	1.717	1.623	1.561	1.516	1.415	1.321	1.189
	0.01	6.763	4.713	3.881	3.414	3.110	2.893	2.730	2.601	2.497	2.411	2.275	2.129	1.971	1.868	1.794	1.629	1.481	1.279
500	0.10	2.716	2.313	2.095	1.956	1.859	1.786	1.729	1.683	1.644	1.612	1.559	1.501	1.435	1.391	1.358	1.282	1.209	1.087
	0.05	3.860	3.014	2.623	2.390	2.232	2.117	2.028	1.957	1.899	1.850	1.772	1.686	1.592	1.528	1.482	1.376	1.275	1.113
	0.01	6.686	4.648	3.821	3.357	3.054	2.838	2.675	2.547	2.443	2.356	2.220	2.075	1.915	1.810	1.735	1.566	1.408	1.164
1000	0.10	2.711	2.308	2.089	1.950	1.853	1.780	1.723	1.676	1.638	1.605	1.552	1.494	1.428	1.383	1.350	1.273	1.197	1.060
	0.05	3.851	3.005	2.614	2.381	2.223	2.108	2.019	1.948	1.889	1.840	1.762	1.676	1.581	1.517	1.471	1.363	1.260	1.078
	0.01	6.660	4.625	3.801	3.338	3.036	2.820	2.657	2.529	2.425	2.339	2.203	2.056	1.897	1.791	1.716	1.544	1.383	1.112
∞	0.10	2.706	2.303	2.084	1.945	1.847	1.774	1.717	1.670	1.632	1.599	1.546	1.487	1.421	1.375	1.342	1.263	1.185	1.000
	0.05	3.841	2.996	2.605	2.372	2.214	2.099	2.010	1.938	1.880	1.831	1.752	1.666	1.571	1.506	1.459	1.350	1.243	1.000
	0.01	6.635	4.605	3.782	3.319	3.017	2.802	2.639	2.511	2.407	2.321	2.185	2.039	1.878	1.773	1.696	1.523	1.358	1.000

(b) Regression and correlation

Table 16. Significance of correlation coefficient, when $\rho = 0$

Values of r significant at α (one-tail) and 2α (two-tail) levels for different number of degrees of freedom ($\nu = n - 2$) where n = number of observations in the sample. **Example:** If $|r| \geq 0.4438$ when $\nu = 18$, H_0 that $\rho = 0$ may be rejected at 5% level of significance ($2\alpha = 0.05$). When testing partial correlation coefficient, $\nu = n - k - 2$, where k = number of variables held constant, e.g. for $r_{12.3}$, $k = 1$.

ν	$\alpha = 0.25$ $2\alpha = 0.5$	$\alpha = 0.15$ $2\alpha = 0.3$	$\alpha = 0.1$ $2\alpha = 0.2$	$\alpha = 0.05$ $2\alpha = 0.1$	$\alpha = 0.025$ $2\alpha = 0.05$	$\alpha = 0.01$ $2\alpha = 0.02$	$\alpha = 0.005$ $2\alpha = 0.01$	$\alpha = 0.001$ $2\alpha = 0.002$	$\alpha = 0.0005$ $2\alpha = 0.001$
1	.7071	.8910	.9511	.9877	.9969	.9995	.9999	1.0000	1.0000
2	.5000	.7000	.8000	.9000	.9500	.9800	.9900	.9980	9990
3	.4040	.5851	.6870	.8054	.8783	.9343	.9587	.9859	.9911
4	.3473	.5112	.6084	.7293	.8114	.8822	.9172	.9633	.9741
5	.3091	.4592	.5509	.6694	.7545	.8329	.8745	.9350	.9509
6	.2811	.4202	.5067	.6215	.7067	.7887	.8343	.9049	.9249
7	.2596	.3896	.4716	.5822	.6664	.7498	.7977	.8751	.8983
8	.2423	.3648	.4428	.5494	.6319	.7155	.7646	.8467	.8721
9	.2281	.3442	.4187	.5214	.6021	.6851	.7348	.8199	.8470
10	.2161	.3267	.3981	.4973	.5760	.6581	.7079	.7950	.8233
11	.2058	.3116	.3802	.4762	.5529	.6339	.6835	.7717	.8010
12	.1968	.2984	.3646	.4575	.5324	.6120	.6614	.7501	.7800
13	.1890	.2868	.3507	.4409	.5140	.5923	.6411	.7301	.7604
14	.1820	.2764	.3383	.4259	.4973	.5742	.6226	.7114	.7419
15	.1757	.2671	.3271	.4124	.4821	.5577	.6055	.6940	.7247
16	.1700	.2587	.3170	.4000	.4683	.5425	.5897	.6777	.7084
17	.1649	.2510	.3077	.3887	.4555	.5285	.5751	.6624	.6932
18	.1602	.2439	.2992	.3783	.4438	.5155	.5614	.6481	.6788
19	.1558	.2375	.2914	.3687	.4329	.5034	.5487	.6346	.6652
20	.1518	.2315	.2841	.3598	.4227	.4921	.5368	.6219	.6524
21	.1481	.2259	.2774	.3515	.4132	.4815	.5256	.6099	.6402
22	.1447	.2207	.2711	.3438	.4044	.4716	.5151	.5986	.6287
23	.1415	.2159	.2653	.3365	.3961	.4622	.5052	.5879	.6178
24	.1384	.2113	.2598	.3297	.3882	.4534	.4958	.5776	.6074
25	.1356	.2071	.2546	.3233	.3809	.4451	.4869	.5679	.5974
26	.1330	.2031	.2497	.3172	.3739	.4372	.4785	.5587	.5880
27	.1305	.1993	.2451	.3115	.3673	.4297	.4705	.5499	.5790
28	.1281	.1957	.2407	.3061	.3610	.4226	.4629	.5415	.5703
29	.1258	.1923	.2366	.3009	.3550	.4158	.4556	.5334	.5620
30	.1237	.1891	.2327	.2960	.3494	.4093	.4487	.5257	.5541
35	.1144	.1751	.2156	.2746	.3246	.3810	.4182	.4916	.5189
40	.1070	.1638	.2018	.2573	.3044	.3578	.3932	.4633	.4896
45	.1008	.1544	.1903	.2429	.2876	.3384	.3721	.4394	.4647
50	.0956	.1465	.1806	.2306	.2732	.3218	.3542	.4188	.4432
60	.0873	.1338	.1650	.2108	.2500	.2948	.3248	.3850	.4079
70	.0808	.1238	.1528	.1954	.2319	.2737	.3017	.3583	.3798
80	.0755	.1158	.1430	.1829	.2172	.2565	.2830	.3364	.3568
90	.0712	.1092	.1348	.1726	.2050	.2422	.2673	.3181	.3375
100	.0675	.1036	.1279	.1638	.1946	.2301	.2540	.3025	.3211
120	.0616	.0946	.1168	.1496	.1779	.2104	.2324	.2771	.2943
150	.0551	.0846	.1045	.1339	.1593	.1886	.2083	.2488	.2643
200	.0477	.0733	.0905	.1161	.1381	.1636	.1809	.2162	.2298
300	.0390	.0598	.0740	.0948	.1129	.1338	.1480	.1771	.1884
500	.0302	.0463	.0573	.0735	.0875	.1038	.1149	.1376	.1464

Table 17. Significance of correlation coefficient, when $\rho \neq 0$

Values of w for different values of r in the relation $w = \frac{1}{2}\log_e\left(\frac{1+r}{1-r}\right)$, or values of ω for different values of ρ in the relation $\omega = \frac{1}{2}\log_e\left(\frac{1+\rho}{1-\rho}\right)$ where r and ρ are sample and population correlation coefficients respectively. The variable w is approximately Normally distributed with mean = ω and standard deviation = $1/\sqrt{n-3}$, where n = sample size. **Example:** If $H_0 \rightarrow \rho = 0.5$, $n = 28$ and $r = 0.746$, then $\omega = 0.5493$, $w = 0.9639$, and, therefore, standardised Normal variable

$z = |w - \omega|\sqrt{n-3} = |0.9639 - 0.5493|\sqrt{28-3} = 2.073$. As $2.073 > 1.96$, H_0 may be rejected at 5% level of significance (two-tailed test). For negative values of r or ρ, w or ω are negative. For partial correlation coefficient, standard deviation = $1/\sqrt{n-k-3}$ where k = number of variables held constant, e.g. for $r_{12.3}$, $k = 1$.

r or $\rho\rightarrow$ ↓	0.000	0.001	0.002	0.003	0.004	0.005	0.006	0.007	0.008	0.009
0.00	0.0000	0.0010	0.0020	0.0030	0.0040	0.0050	0.0060	0.0070	0.0080	0.0090
0.01	0.0100	0.0110	0.0120	0.0130	0.0140	0.0150	0.0160	0.0170	0.0180	0.0190
0.02	0.0200	0.0210	0.0220	0.0230	0.0240	0.0250	0.0260	0.0270	0.0280	0.0290
0.03	0.0300	0.0310	0.0320	0.0330	0.0340	0.0350	0.0360	0.0370	0.0380	0.0390
0.04	0.0400	0.0410	0.0420	0.0430	0.0440	0.0450	0.0460	0.0470	0.0480	0.0490
0.05	0.0500	0.0510	0.0520	0.0530	0.0541	0.0551	0.0561	0.0571	0.0581	0.0591
0.06	0.0601	0.0611	0.0621	0.0631	0.0641	0.0651	0.0661	0.0671	0.0681	0.0691
0.07	0.0701	0.0711	0.0721	0.0731	0.0741	0.0751	0.0761	0.0772	0.0782	0.0792
0.08	0.0802	0.0812	0.0822	0.0832	0.0842	0.0852	0.0862	0.0872	0.0882	0.0892
0.09	0.0902	0.0913	0.0923	0.0933	0.0943	0.0953	0.0963	0.0973	0.0983	0.0993
0.10	0.1003	0.1013	0.1024	0.1034	0.1044	0.1054	0.1064	0.1074	0.1084	0.1094
0.11	0.1104	0.1115	0.1125	0.1135	0.1145	0.1155	0.1165	0.1175	0.1186	0.1196
0.12	0.1206	0.1216	0.1226	0.1236	0.1246	0.1257	0.1267	0.1277	0.1287	0.1297
0.13	0.1307	0.1318	0.1328	0.1338	0.1348	0.1358	0.1368	0.1379	0.1389	0.1399
0.14	0.1409	0.1419	0.1430	0.1440	0.1450	0.1460	0.1471	0.1481	0.1491	0.1501
0.15	0.1511	0.1522	0.1532	0.1542	0.1552	0.1563	0.1573	0.1583	0.1593	0.1604
0.16	0.1614	0.1624	0.1634	0.1645	0.1655	0.1665	0.1676	0.1686	0.1696	0.1706
0.17	0.1717	0.1727	0.1737	0.1748	0.1758	0.1768	0.1779	0.1789	0.1799	0.1809
0.18	0.1820	0.1830	0.1841	0.1851	0.1861	0.1872	0.1882	0.1892	0.1903	0.1913
0.19	0.1923	0.1934	0.1944	0.1955	0.1965	0.1975	0.1986	0.1996	0.2007	0.2017
0.20	0.2027	0.2038	0.2048	0.2059	0.2069	0.2079	0.2090	0.2100	0.2111	0.2121
0.21	0.2132	0.2142	0.2153	0.2163	0.2174	0.2184	0.2195	0.2205	0.2216	0.2226
0.22	0.2237	0.2247	0.2258	0.2268	0.2279	0.2289	0.2300	0.2310	0.2321	0.2331
0.23	0.2342	0.2352	0.2363	0.2374	0.2384	0.2395	0.2405	0.2416	0.2427	0.2437
0.24	0.2448	0.2458	0.2469	0.2480	0.2490	0.2501	0.2512	0.2522	0.2533	0.2543
0.25	0.2554	0.2565	0.2575	0.2586	0.2597	0.2608	0.2618	0.2629	0.2640	0.2650
0.26	0.2661	0.2672	0.2683	0.2693	0.2704	0.2715	0.2726	0.2736	0.2747	0.2758
0.27	0.2769	0.2779	0.2790	0.2801	0.2812	0.2823	0.2833	0.2844	0.2855	0.2866
0.28	0.2877	0.2888	0.2899	0.2909	0.2920	0.2931	0.2942	0.2953	0.2964	0.2975
0.29	0.2986	0.2997	0.3008	0.3018	0.3029	0.3040	0.3051	0.3062	0.3073	0.3084
0.30	0.3095	0.3106	0.3117	0.3128	0.3139	0.3150	0.3161	0.3172	0.3183	0.3194
0.31	0.3205	0.3217	0.3228	0.3239	0.3250	0.3261	0.3272	0.3283	0.3294	0.3305
0.32	0.3316	0.3328	0.3339	0.3350	0.3361	0.3372	0.3383	0.3395	0.3406	0.3417
0.33	0.3428	0.3440	0.3451	0.3462	0.3473	0.3484	0.3496	0.3507	0.3518	0.3530
0.34	0.3541	0.3552	0.3564	0.3575	0.3586	0.3598	0.3609	0.3620	0.3632	0.3643
0.35	0.3654	0.3666	0.3677	0.3689	0.3700	0.3712	0.3723	0.3734	0.3746	0.3757
0.36	0.3769	0.3780	0.3792	0.3803	0.3815	0.3826	0.3838	0.3850	0.3861	0.3873
0.37	0.3884	0.3896	0.3907	0.3919	0.3931	0.3942	0.3954	0.3966	0.3977	0.3989
0.38	0.4001	0.4012	0.4024	0.4036	0.4047	0.4059	0.4071	0.4083	0.4094	0.4106
0.39	0.4118	0.4130	0.4142	0.4153	0.4165	0.4177	0.4189	0.4201	0.4213	0.4225
0.40	0.4236	0.4248	0.4260	0.4272	0.4284	0.4296	0.4308	0.4320	0.4332	0.4344
0.41	0.4356	0.4368	0.4380	0.4392	0.4404	0.4416	0.4428	0.4441	0.4453	0.4465
0.42	0.4477	0.4489	0.4501	0.4513	0.4526	0.4538	0.4550	0.4562	0.4574	0.4587
0.43	0.4599	0.4611	0.4624	0.4636	0.4648	0.4660	0.4673	0.4685	0.4698	0.4710
0.44	0.4722	0.4735	0.4747	0.4760	0.4772	0.4784	0.4797	0.4809	0.4822	0.4834
0.45	0.4847	0.4860	0.4872	0.4885	0.4897	0.4910	0.4922	0.4935	0.4948	0.4960
0.46	0.4973	0.4986	0.4999	0.5011	0.5024	0.5037	0.5049	0.5062	0.5075	0.5088
0.47	0.5101	0.5114	0.5126	0.5139	0.5152	0.5165	0.5178	0.5191	0.5204	0.5217
0.48	0.5230	0.5243	0.5256	0.5269	0.5282	0.5295	0.5308	0.5321	0.5334	0.5347
0.49	0.5361	0.5374	0.5387	0.5400	0.5413	0.5427	0.5440	0.5453	0.5466	0.5480

Table 17 – *continued*

r or ρ→ ↓	0.000	0.001	0.002	0.003	0.004	0.005	0.006	0.007	0.008	0.009
0.50	0.5493	0.5506	0.5520	0.5533	0.5547	0.5560	0.5573	0.5587	0.5600	0.5614
0.51	0.5627	0.5641	0.5654	0.5668	0.5682	0.5695	0.5709	0.5722	0.5736	0.5750
0.52	0.5763	0.5777	0.5791	0.5805	0.5818	0.5832	0.5846	0.5860	0.5874	0.5888
0.53	0.5901	0.5915	0.5929	0.5943	0.5957	0.5971	0.5985	0.5999	0.6013	0.6027
0.54	0.6042	0.6056	0.6070	0.6084	0.6098	0.6112	0.6127	0.6141	0.6155	0.6169
0.55	0.6184	0.6198	0.6213	0.6227	0.6241	0.6256	0.6270	0.6285	0.6299	0.6314
0.56	0.6328	0.6343	0.6358	0.6372	0.6387	0.6401	0.6416	0.6431	0.6446	0.6460
0.57	0.6475	0.6490	0.6505	0.6520	0.6535	0.6550	0.6565	0.6580	0.6595	0.6610
0.58	0.6625	0.6640	0.6655	0.6670	0.6685	0.6700	0.6716	0.6731	0.6746	0.6761
0.59	0.6777	0.6792	0.6807	0.6823	0.6838	0.6854	0.6869	0.6885	0.6900	0.6916
0.60	0.6931	0.6947	0.6963	0.6978	0.6994	0.7010	0.7026	0.7042	0.7057	0.7073
0.61	0.7089	0.7105	0.7121	0.7137	0.7153	0.7169	0.7185	0.7201	0.7218	0.7234
0.62	0.7250	0.7266	0.7283	0.7299	0.7315	0.7332	0.7348	0.7365	0.7381	0.7398
0.63	0.7414	0.7431	0.7447	0.7464	0.7481	0.7498	0.7514	0.7531	0.7548	0.7565
0.64	0.7582	0.7599	0.7616	0.7633	0.7650	0.7667	0.7684	0.7701	0.7718	0.7736
0.65	0.7753	0.7770	0.7788	0.7805	0.7823	0.7840	0.7858	0.7875	0.7893	0.7910
0.66	0.7928	0.7946	0.7964	0.7981	0.7999	0.8017	0.8035	0.8053	0.8071	0.8089
0.67	0.8107	0.8126	0.8144	0.8162	0.8180	0.8199	0.8217	0.8236	0.8254	0.8273
0.68	0.8291	0.8310	0.8328	0.8347	0.8366	0.8385	0.8404	0.8423	0.8441	0.8460
0.69	0.8480	0.8499	0.8518	0.8537	0.8556	0.8576	0.8595	0.8614	0.8634	0.8653
0.70	0.8673	0.8693	0.8712	0.8732	0.8752	0.8772	0.8792	0.8812	0.8832	0.8852
0.71	0.8872	0.8892	0.8912	0.8933	0.8953	0.8973	0.8994	0.9014	0.9035	0.9056
0.72	0.9076	0.9097	0.9118	0.9139	0.9160	0.9181	0.9202	0.9223	0.9245	0.9266
0.73	0.9287	0.9309	0.9330	0.9352	0.9373	0.9395	0.9417	0.9439	0.9461	0.9483
0.74	0.9505	0.9527	0.9549	0.9571	0.9594	0.9616	0.9639	0.9661	0.9684	0.9707
0.75	0.9730	0.9752	0.9775	0.9798	0.9822	0.9845	0.9868	0.9892	0.9915	0.9939
0.76	0.9962	0.9986	1.0010	1.0034	1.0058	1.0082	1.0106	1.0130	1.0154	1.0179
0.77	1.0203	1.0228	1.0253	1.0277	1.0302	1.0327	1.0352	1.0378	1.0403	1.0428
0.78	1.0454	1.0479	1.0505	1.0531	1.0557	1.0583	1.0609	1.0635	1.0661	1.0688
0.79	1.0714	1.0741	1.0768	1.0795	1.0822	1.0849	1.0876	1.0903	1.0931	1.0958
0.80	1.0986	1.1014	1.1042	1.1070	1.1098	1.1127	1.1155	1.1184	1.1212	1.1241
0.81	1.1270	1.1299	1.1329	1.1358	1.1388	1.1417	1.1447	1.1477	1.1507	1.1538
0.82	1.1568	1.1599	1.1630	1.1660	1.1692	1.1723	1.1754	1.1786	1.1817	1.1849
0.83	1.1881	1.1914	1.1946	1.1979	1.2011	1.2044	1.2077	1.2111	1.2144	1.2178
0.84	1.2212	1.2246	1.2280	1.2315	1.2349	1.2384	1.2419	1.2454	1.2490	1.2526
0.85	1.2562	1.2598	1.2634	1.2671	1.2707	1.2745	1.2782	1.2819	1.2857	1.2895
0.86	1.2933	1.2972	1.3011	1.3050	1.3089	1.3129	1.3169	1.3209	1.3249	1.3290
0.87	1.3331	1.3372	1.3414	1.3456	1.3498	1.3540	1.3583	1.3626	1.3670	1.3714
0.88	1.3758	1.3802	1.3847	1.3892	1.3938	1.3984	1.4030	1.4077	1.4124	1.4171
0.89	1.4219	1.4268	1.4316	1.4365	1.4415	1.4465	1.4516	1.4566	1.4618	1.4670
0.90	1.4722	1.4775	1.4828	1.4882	1.4937	1.4992	1.5047	1.5103	1.5160	1.5217
0.91	1.5275	1.5334	1.5393	1.5453	1.5513	1.5574	1.5636	1.5698	1.5762	1.5826
0.92	1.5890	1.5956	1.6022	1.6089	1.6157	1.6226	1.6296	1.6366	1.6438	1.6510
0.93	1.6584	1.6658	1.6734	1.6811	1.6888	1.6967	1.7047	1.7129	1.7211	1.7295
0.94	1.7380	1.7467	1.7555	1.7645	1.7736	1.7828	1.7923	1.8019	1.8117	1.8216
0.95	1.8318	1.8421	1.8527	1.8635	1.8745	1.8857	1.8972	1.9090	1.9210	1.9333
0.96	1.9459	1.9588	1.9721	1.9857	1.9996	2.0139	2.0287	2.0439	2.0595	2.0756
0.97	2.0923	2.1095	2.1273	2.1457	2.1649	2.1847	2.2054	2.2269	2.2494	2.2729
0.98	2.2976	2.3235	2.3507	2.3796	2.4101	2.4427	2.4774	2.5147	2.5550	2.5987
0.99	2.6467	2.6996	2.7587	2.8257	2.9031	2.9945	3.1063	3.2504	3.4534	3.8002

Table 18. Significance of multiple correlation coefficient

Values of R^2 significant at 10%, 5% and 1% for different number of degrees of freedom, ν, and number of explanatory (independent) variables, k, where $\nu = n - k - 1$ and n = sample size. **Example:** In regression of Y on X and Z based on $n = 20$, $k = 2$, $\nu = 20 - 2 - 1 = 17$, and 5% value of $R^2 = 0.2970$, i.e. sample values of $R^2 \geqslant 0.2970$ are significantly different from zero at 5% level.

	$k = 1$			$k = 2$			$k = 3$			$k = 4$			$k = 5$		
ν	10%	5%	1%	10%	5%	1%	10%	5%	1%	10%	5%	1%	10%	5%	1%
1	.9755	.9938	.9998	.9900	.9975	.9999	.9938	.9985	.9999	.9955	.9989	1.0000	.9965	.9991	1.0000
2	.8100	.9025	.9801	.9000	.9500	.9900	.9322	.9664	.9933	.9487	.9747	.9950	.9587	.9797	.9960
3	.6486	.7715	.9192	.7846	.8643	.9536	.8435	.9027	.9672	.8769	.9240	.9745	.8985	.9376	.9792
4	.5319	.6584	.8413	.8838	.7764	.9000	.7586	.8318	.9260	.8042	.8640	.9411	.8351	.8866	.9510
5	.4481	.5693	.7648	.6019	.6983	.8415	.6847	.7645	.8786	.7380	.8060	.9011	.7754	.8347	.9164
6	.3862	.4995	.6961	.5358	.6316	.7846	.6218	.7040	.8302	.6795	.7514	.8591	.7214	.7852	.8793
7	.3390	.4441	.6303	.4821	.5751	.7317	.5685	.6507	.7836	.6285	.7019	.8176	.6732	.7394	.8420
8	.3018	.3993	.5846	.4377	.5271	.6838	.5230	.6039	.7400	.5839	.6574	.7779	.6302	.6974	.8056
9	.2719	.3625	.5399	.4005	.4861	.6406	.4839	.5628	.6998	.5448	.6175	.7405	.5919	.6592	.7709
10	.2473	.3318	.5011	.3690	.4507	.6019	.4500	.5266	.6628	.5103	.5818	.7057	.5577	.6245	.7381
11	.2267	.3057	.4672	.3421	.4200	.5671	.4205	.4945	.6290	.4798	.5497	.6733	.5270	.5929	.7073
12	.2093	.2835	.4374	.3187	.3930	.5358	.3944	.4660	.5981	.4526	.5207	.6434	.4994	.5641	.6785
13	.1944	.2642	.4111	.2983	.3693	.5076	.3714	.4404	.5698	.4282	.4945	.6156	.4744	.5378	.6516
14	.1814	.2473	.3876	.2803	.3482	.4821	.3509	.4174	.5438	.4062	.4707	.5899	.4517	.5137	.6264
15	.1700	.2325	.3666	.2644	.3293	.4588	.3324	.3967	.5200	.3864	.4490	.5661	.4311	.4916	.6029
16	.1600	.2193	.3478	.2501	.3123	.4377	.3158	.3778	.4981	.3684	.4291	.5440	.4122	.4713	.5810
17	.1511	.2075	.3307	.2373	.2970	.4183	.3008	.3607	.4778	.3519	.4109	.5235	.3948	.4525	.5605
18	.1431	.1969	.3152	.2257	.2831	.4005	.2871	.3450	.4591	.3368	.3942	.5044	.3789	.4351	.5413
19	.1360	.1874	.3011	.2152	.2705	.3842	.2746	.3306	.4417	.3230	.3787	.4865	.3641	.4190	.5233
20	.1295	.1787	.2882	.2057	.2589	.3690	.2631	.3173	.4255	.3102	.3644	.4698	.3505	.4040	.5063
21	.1236	.1708	.2763	.1969	.2482	.3551	.2525	.3050	.4105	.2984	.3511	.4542	.3378	.3900	.4904
22	.1182	.1635	.2653	.1889	.2384	.3421	.2428	.2937	.3964	.2875	.3387	.4395	.3260	.3769	.4754
23	.1132	.1569	.2552	.1815	.2293	.3300	.2337	.2831	.3833	.2773	.3271	.4258	.3150	.3646	.4613
24	.1087	.1507	.2458	.1746	.2209	.3187	.2254	.2733	.3710	.2678	.3163	.4128	.3047	.3532	.4480
25	.1045	.1451	.2371	.1682	.2131	.3082	.2176	.2641	.3594	.2590	.3062	.4006	.2950	.3424	.4353
26	.1006	.1398	.2290	.1623	.2058	.2983	.2103	.2556	.3485	.2507	.2967	.3891	.2859	.3322	.4234
27	.0970	.1349	.2214	.1568	.1990	.2890	.2034	.2475	.3383	.2429	.2878	.3782	.2774	.3226	.4121
28	.0937	.1303	.2143	.1517	.1926	.2803	.1971	.2400	.3286	.2356	.2794	.3679	.2694	.3136	.4013
29	.0905	.1261	.2076	.1468	.1867	.2721	.1911	.2328	.3195	.2287	.2715	.3581	.2618	.3050	.3911
30	.0876	.1221	.2013	.1423	.1810	.2644	.1854	.2261	.3108	.2222	.2640	.3488	.2546	.2969	.3814
35	.0754	.1053	.1749	.1233	.1573	.2314	.1615	.1977	.2737	.1945	.2319	.3088	.2239	.2620	.3391
40	.0662	.0927	.1546	.1087	.1391	.2057	.1431	.1755	.2444	.1729	.2067	.2768	.1997	.2344	.3052
45	.0590	.0827	.1385	.0973	.1247	.1851	.1284	.1578	.2207	.1557	.1865	.2509	.1803	.2121	.2774
50	.0532	.0747	.1254	.0880	.1129	.1682	.1165	.1434	.2013	.1415	.1698	.2293	.1643	.1936	.2542
60	.0445	.0625	.1055	.0739	.0950	.1423	.0982	.1212	.1710	.1198	.1441	.1957	.1395	.1648	.2177
70	.0382	.0538	.0910	.0637	.0820	.1233	.0849	.1049	.1487	.1038	.1251	.1706	.1212	.1435	.1903
80	.0335	.0472	.0801	.0559	.0722	.1087	.0747	.0925	.1315	.0916	.1106	.1512	.1072	.1271	.1690
90	.0298	.0420	.0714	.0499	.0644	.0973	.0667	.0827	.1178	.0819	.0990	.1358	.0960	.1140	.1520
100	.0268	.0379	.0645	.0450	.0582	.0880	.0603	.0748	.1068	.0741	.0897	.1232	.0870	.1034	.1381
120	.0224	.0316	.0540	.0376	.0487	.0739	.0506	.0628	.0899	.0623	.0754	.1039	.0732	.0871	.1168
150	.0179	.0254	.0434	.0302	.0392	.0596	.0407	.0506	.0726	.0502	.0609	.0842	.0591	.0705	.0948
200	.0135	.0191	.0327	.0228	.0295	.0450	.0307	.0382	.0550	.0380	.0461	.0639	.0448	.0535	.0721
500	.0054	.0077	.0132	.0092	.0119	.0183	.0124	.0155	.0224	.0154	.0188	.0262	.0182	.0218	.0296
1000	.0027	.0038	.0066	.0046	.0060	.0092	.0062	.0078	.0113	.0077	.0094	.0132	.0092	.0110	.0150

Table 19. Durbin–Watson d-statistic

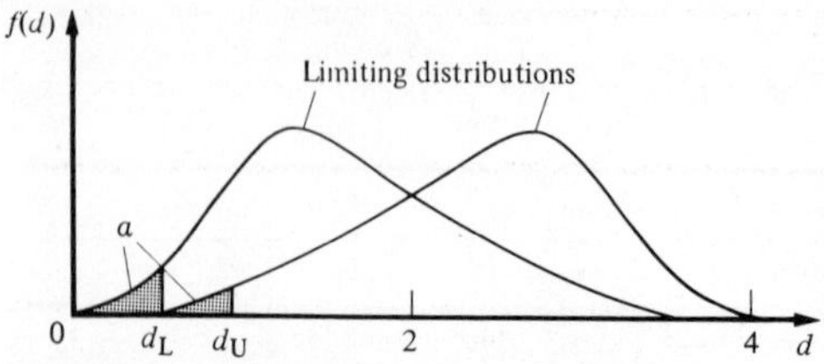

Lower (d_L) and upper (d_U) points of the limiting distributions of d for different sample sizes (n), number of explanatory variables (k) and significance levels (α). **Example:** In regression of Y on X and Z (i.e. for $k = 2$) when $n = 20$ and $\alpha = 0.05$, $d_L = 1.10$ and $d_U = 1.54$. If sample $d < 1.10$ the hypothesis of no autocorrelation (i.e. $\rho = 0$) can be rejected; if $d > 1.54$, $\rho = 0$ cannot be rejected; if $1.10 < d < 1.54$ the test is inconclusive. For negative autocorrelation find $4 - d$ and test as above.

n	α	$k = 1$		$k = 2$		$k = 3$		$k = 4$		$k = 5$	
		d_L	d_U	d_L	d_U	d_L	d_U	d_L	d_U	d_L	d_U
15	.05	1.08	1.36	0.95	1.54	0.82	1.75	0.69	1.97	0.56	2.21
	.025	0.95	1.23	0.83	1.40	0.71	1.61	0.59	1.84	0.48	2.09
	.01	0.81	1.07	0.70	1.25	0.59	1.46	0.49	1.70	0.39	1.96
16	.05	1.10	1.37	0.98	1.54	0.86	1.73	0.74	1.93	0.62	2.15
	.025	0.98	1.24	0.86	1.40	0.75	1.59	0.64	1.80	0.53	2.03
	.01	0.84	1.09	0.74	1.25	0.63	1.44	0.53	1.66	0.44	1.90
17	.05	1.13	1.38	1.02	1.54	0.90	1.71	0.78	1.90	0.67	2.10
	.025	1.01	1.25	0.90	1.40	0.79	1.58	0.68	1.77	0.57	1.98
	.01	0.87	1.10	0.77	1.25	0.67	1.43	0.57	1.63	0.48	1.85
18	.05	1.16	1.39	1.05	1.53	0.93	1.69	0.82	1.87	0.71	2.06
	.025	1.03	1.26	0.93	1.40	0.82	1.56	0.72	1.74	0.62	1.93
	.01	0.90	1.12	0.80	1.26	0.71	1.42	0.61	1.60	0.52	1.80
19	.05	1.18	1.40	1.08	1.53	0.97	1.68	0.86	1.85	0.75	2.02
	.025	1.06	1.28	0.96	1.41	0.86	1.55	0.76	1.72	0.66	1.90
	.01	0.93	1.13	0.83	1.26	0.74	1.41	0.65	1.58	0.56	1.77
20	.05	1.20	1.41	1.10	1.54	1.00	1.68	0.90	1.83	0.79	1.99
	.025	1.08	1.28	0.99	1.41	0.89	1.55	0.79	1.70	0.70	1.87
	.01	0.95	1.15	0.86	1.27	0.77	1.41	0.68	1.57	0.60	1.74
21	.05	1.22	1.42	1.13	1.54	1.03	1.67	0.93	1.81	0.83	1.96
	.025	1.10	1.30	1.01	1.41	0.92	1.54	0.83	1.69	0.73	1.84
	.01	0.97	1.16	0.89	1.27	0.80	1.41	0.72	1.55	0.63	1.71
22	.05	1.24	1.43	1.15	1.54	1.05	1.66	0.96	1.80	0.86	1.94
	.025	1.12	1.31	1.04	1.42	0.95	1.54	0.86	1.68	0.77	1.82
	.01	1.00	1.17	0.91	1.28	0.83	1.40	0.75	1.54	0.66	1.69
23	.05	1.26	1.44	1.17	1.54	1.08	1.66	0.99	1.79	0.90	1.92
	.025	1.14	1.32	1.06	1.42	0.97	1.54	0.89	1.67	0.80	1.80
	.01	1.02	1.19	0.94	1.29	0.86	1.40	0.77	1.53	0.70	1.67
24	.05	1.27	1.45	1.19	1.55	1.10	1.66	1.01	1.78	0.93	1.90
	.025	1.16	1.33	1.08	1.43	1.00	1.54	0.91	1.66	0.83	1.79
	.01	1.04	1.20	0.96	1.30	0.88	1.41	0.80	1.53	0.72	1.66
25	.05	1.29	1.45	1.21	1.55	1.12	1.66	1.04	1.77	0.95	1.89
	.025	1.18	1.34	1.10	1.43	1.02	1.54	0.94	1.65	0.86	1.77
	.01	1.05	1.21	0.98	1.30	0.90	1.41	0.83	1.52	0.75	1.65
26	.05	1.30	1.46	1.22	1.55	1.14	1.65	1.06	1.76	0.98	1.88
	.025	1.19	1.35	1.12	1.44	1.04	1.54	0.96	1.65	0.88	1.76
	.01	1.07	1.22	1.00	1.31	0.93	1.41	0.85	1.52	0.78	1.64
27	.05	1.32	1.47	1.24	1.56	1.16	1.65	1.08	1.76	1.01	1.86
	.025	1.21	1.36	1.13	1.44	1.06	1.54	0.99	1.64	0.91	1.75
	.01	1.09	1.23	1.02	1.32	0.95	1.41	0.88	1.51	0.81	1.63
28	.05	1.33	1.48	1.26	1.56	1.18	1.65	1.10	1.75	1.03	1.85
	.025	1.22	1.37	1.15	1.45	1.08	1.54	1.01	1.64	0.93	1.74
	.01	1.10	1.24	1.04	1.32	0.97	1.41	0.90	1.51	0.83	1.62

Table 19 – *continued*

n	α	$k = 1$		$k = 2$		$k = 3$		$k = 4$		$k = 5$	
		d_L	d_U	d_L	d_U	d_L	d_U	d_L	d_U	d_L	d_U
29	.05	1.34	1.48	1.27	1.56	1.20	1.65	1.12	1.74	1.05	1.84
	.025	1.24	1.38	1.17	1.45	1.10	1.54	1.03	1.63	0.96	1.73
	.01	1.12	1.25	1.05	1.33	0.99	1.42	0.92	1.51	0.85	1.61
30	.05	1.35	1.49	1.28	1.57	1.21	1.65	1.14	1.74	1.07	1.83
	.025	1.25	1.38	1.18	1.46	1.12	1.54	1.05	1.63	0.98	1.73
	.01	1.13	1.26	1.07	1.34	1.01	1.42	0.94	1.51	0.88	1.61
31	.05	1.36	1.50	1.30	1.57	1.23	1.65	1.16	1.74	1.09	1.83
	.025	1.26	1.39	1.20	1.47	1.13	1.55	1.07	1.63	1.00	1.72
	.01	1.15	1.27	1.08	1.34	1.02	1.42	0.96	1.51	0.90	1.60
32	.05	1.37	1.50	1.31	1.57	1.24	1.65	1.18	1.73	1.11	1.82
	.025	1.27	1.40	1.21	1.47	1.15	1.55	1.08	1.63	1.02	1.71
	.01	1.16	1.28	1.10	1.35	1.04	1.43	0.98	1.51	0.92	1.60
33	.05	1.38	1.51	1.32	1.58	1.26	1.65	1.19	1.73	1.13	1.81
	.025	1.28	1.41	1.22	1.48	1.16	1.55	1.10	1.63	1.04	1.71
	.01	1.17	1.29	1.11	1.36	1.05	1.43	1.00	1.51	0.94	1.59
34	.05	1.39	1.51	1.33	1.58	1.27	1.65	1.21	1.73	1.15	1.81
	.025	1.29	1.41	1.24	1.48	1.17	1.55	1.12	1.63	1.06	1.70
	.01	1.18	1.30	1.13	1.36	1.07	1.43	1.01	1.51	0.95	1.59
35	.05	1.40	1.52	1.34	1.58	1.28	1.65	1.22	1.73	1.16	1.80
	.025	1.30	1.42	1.25	1.48	1.19	1.55	1.13	1.63	1.07	1.70
	.01	1.19	1.31	1.14	1.37	1.08	1.44	1.03	1.51	0.97	1.59
36	.05	1.41	1.52	1.35	1.59	1.29	1.65	1.24	1.73	1.18	1.80
	.025	1.31	1.43	1.26	1.49	1.20	1.56	1.15	1.63	1.09	1.70
	.01	1.21	1.32	1.15	1.38	1.10	1.44	1.04	1.51	0.99	1.59
37	.05	1.42	1.53	1.36	1.59	1.31	1.66	1.25	1.72	1.19	1.80
	.025	1.32	1.43	1.27	1.49	1.21	1.56	1.16	1.62	1.10	1.70
	.01	1.22	1.32	1.16	1.38	1.11	1.45	1.06	1.51	1.00	1.59
38	.05	1.43	1.54	1.37	1.59	1.32	1.66	1.26	1.72	1.21	1.79
	.025	1.33	1.44	1.28	1.50	1.23	1.56	1.17	1.62	1.12	1.70
	.01	1.23	1.33	1.18	1.39	1.12	1.45	1.07	1.52	1.02	1.58
39	.05	1.43	1.54	1.38	1.60	1.33	1.66	1.27	1.72	1.22	1.79
	.025	1.34	1.44	1.29	1.50	1.24	1.56	1.19	1.63	1.13	1.69
	.01	1.24	1.34	1.19	1.39	1.14	1.45	1.09	1.52	1.03	1.58
40	.05	1.44	1.54	1.39	1.60	1.34	1.66	1.29	1.72	1.23	1.79
	.025	1.35	1.45	1.30	1.51	1.25	1.57	1.20	1.63	1.15	1.69
	.01	1.25	1.34	1.20	1.40	1.15	1.46	1.10	1.52	1.05	1.58
45	.05	1.48	1.57	1.43	1.62	1.38	1.67	1.34	1.72	1.29	1.78
	.025	1.39	1.48	1.34	1.53	1.30	1.58	1.25	1.63	1.21	1.69
	.01	1.29	1.38	1.24	1.42	1.20	1.48	1.16	1.53	1.11	1.58
50	.05	1.50	1.59	1.46	1.63	1.42	1.67	1.38	1.72	1.34	1.77
	.025	1.42	1.50	1.38	1.54	1.34	1.59	1.30	1.64	1.26	1.69
	.01	1.32	1.40	1.28	1.45	1.24	1.49	1.20	1.54	1.16	1.59
55	.05	1.53	1.60	1.49	1.64	1.45	1.68	1.41	1.72	1.38	1.77
	.025	1.45	1.52	1.41	1.56	1.37	1.60	1.33	1.64	1.30	1.69
	.01	1.36	1.43	1.32	1.47	1.28	1.51	1.25	1.55	1.21	1.59
60	.05	1.55	1.62	1.51	1.65	1.48	1.69	1.44	1.73	1.41	1.77
	.025	1.47	1.54	1.44	1.57	1.40	1.61	1.37	1.65	1.33	1.69
	.01	1.38	1.45	1.35	1.48	1.32	1.52	1.28	1.56	1.25	1.60
65	.05	1.57	1.63	1.54	1.66	1.50	1.70	1.47	1.73	1.44	1.77
	.025	1.49	1.55	1.46	1.59	1.43	1.62	1.40	1.66	1.36	1.69
	.01	1.41	1.47	1.38	1.50	1.35	1.53	1.31	1.57	1.28	1.61

Table 19 – *continued*

n	α	$k = 1$		$k = 2$		$k = 3$		$k = 4$		$k = 5$	
		d_L	d_U	d_L	d_U	d_L	d_U	d_L	d_U	d_L	d_U
70	.05	1.58	1.64	1.55	1.67	1.52	1.70	1.49	1.74	1.46	1.77
	.025	1.51	1.57	1.48	1.60	1.45	1.63	1.42	1.66	1.39	1.70
	.01	1.43	1.49	1.40	1.52	1.37	1.55	1.34	1.58	1.31	1.61
75	.05	1.60	1.65	1.57	1.68	1.54	1.71	1.51	1.74	1.49	1.77
	.025	1.53	1.58	1.50	1.61	1.47	1.64	1.45	1.67	1.42	1.70
	.01	1.45	1.50	1.42	1.53	1.39	1.56	1.37	1.59	1.34	1.62
80	.05	1.61	1.66	1.59	1.69	1.56	1.72	1.53	1.74	1.51	1.77
	.025	1.54	1.59	1.52	1.62	1.49	1.65	1.47	1.67	1.44	1.70
	.01	1.47	1.52	1.44	1.54	1.42	1.57	1.39	1.60	1.36	1.62
85	.05	1.62	1.67	1.60	1.70	1.57	1.72	1.55	1.75	1.52	1.77
	.025	1.56	1.60	1.53	1.63	1.51	1.65	1.49	1.68	1.46	1.71
	.01	1.48	1.53	1.46	1.55	1.43	1.58	1.41	1.60	1.39	1.63
90	.05	1.63	1.68	1.61	1.70	1.59	1.73	1.57	1.75	1.54	1.78
	.025	1.57	1.61	1.55	1.64	1.53	1.66	1.50	1.69	1.48	1.71
	.01	1.50	1.54	1.47	1.56	1.45	1.59	1.43	1.61	1.41	1.64
95	.05	1.64	1.69	1.62	1.71	1.60	1.73	1.58	1.75	1.56	1.78
	.025	1.58	1.62	1.56	1.65	1.54	1.67	1.52	1.69	1.50	1.71
	.01	1.51	1.55	1.49	1.57	1.47	1.60	1.45	1.62	1.42	1.64
100	.05	1.65	1.69	1.63	1.72	1.61	1.74	1.59	1.76	1.57	1.78
	.025	1.59	1.63	1.57	1.65	1.55	1.67	1.53	1.70	1.51	1.72
	.01	1.52	1.56	1.50	1.58	1.48	1.60	1.46	1.63	1.44	1.65

(c) Nonparametric statistics

Table 20. Kolmogorov–Smirnov one-sample test

Critical values of D = maximum $|F_0(X) - S_n(X)|$, where $F_0(X)$ = expected (theoretical) cumulative frequency distribution ($\leqslant X$) and $S_n(X)$ = observed cumulative frequency distribution ($\leqslant X$) of a random sample of n observations. **Example:** For a sample of $n = 25$, observed values of $D \geqslant 0.26404$ are significant at $\alpha = 0.05$ level (two-tailed test). For a one-tailed test the sign of D has to be noted and α halved.

Sample size n	Level of significance for D					Sample size n	Level of significance for D				
	0.2	0.1	0.05	0.02	0.01		0.2	0.1	0.05	0.02	0.01
1	.90000	.95000	.97500	.99000	.99500	51	.14697	.16796	.18659	.20864	.22386
2	.68377	.77639	.84189	.90000	.92929	52	.14558	.16637	.18482	.20667	.22174
3	.56481	.63604	.70760	.78456	.82900	53	.14423	.16483	.18311	.20475	.21968
4	.49265	.56522	.62394	.68887	.73424	54	.14292	.16332	.18144	.20289	.21768
5	.44698	.50945	.56328	.62718	.66853	55	.14164	.16186	.17981	.20107	.21574
6	.41037	.46799	.51926	.57741	.61661	56	.14040	.16044	.17823	.19930	.21384
7	.38148	.43607	.48342	.53844	.57581	57	.13919	.15906	.17669	.19758	.21199
8	.35831	.40962	.45427	.50654	.54179	58	.13801	.15771	.17519	.19590	.21019
9	.33910	.38746	.43001	.47960	.51332	59	.13686	.15639	.17373	.19427	.20844
10	.32260	.36866	.40925	.45662	.48893	60	.13573	.15511	.17231	.19267	.20673
11	.30829	.35242	.39122	.43670	.46770	61	.13464	.15385	.17091	.19112	.20506
12	.29577	.33815	.37543	.41918	.44905	62	.13357	.15263	.16956	.18960	.20343
13	.28470	.32549	.36143	.40362	.43247	63	.13253	.15144	.16823	.18812	.20184
14	.27481	.31417	.34890	.38970	.41762	64	.13151	.15027	.16693	.18667	.20029
15	.26588	.30397	.33760	.37713	.40420	65	.13052	.14913	.16567	.18525	.19877
16	.25778	.29472	.32733	.36571	.39201	66	.12954	.14802	.16443	.18387	.19729
17	.25039	.28627	.31796	.35528	.38086	67	.12859	.14693	.16322	.18252	.19584
18	.24360	.27851	.30936	.34569	.37062	68	.12766	.14587	.16204	.18119	.19442
19	.23735	.27136	.30143	.33685	.36117	69	.12675	.14483	.16088	.17990	.19303
20	.23156	.26473	.29408	.32866	.35241	70	.12586	.14381	.15975	.17863	.19167
21	.22617	.25858	.28724	.32104	.34427	71	.12499	.14281	.15864	.17739	.19034
22	.22115	.25283	.28087	.31394	.33666	72	.12413	.14183	.15755	.17618	.18903
23	.21645	.24746	.27490	.30728	.32954	73	.12329	.14087	.15649	.17498	.18776
24	.21205	.24242	.26931	.30104	.32286	74	.12247	.13993	.15544	.17382	.18650
25	.20790	.23768	.26404	.29516	.31657	75	.12167	.13901	.15442	.17268	.18528
26	.20399	.23320	.25907	.28962	.31064	76	.12088	.13811	.15342	.17155	.18408
27	.20030	.22898	.25438	.28438	.30502	77	.12011	.13723	.15244	.17045	.18290
28	.19680	.22497	.24993	.27942	.29971	78	.11935	.13636	.15147	.16938	.18174
29	.19348	.22117	.24571	.27471	.29466	79	.11860	.13551	.15052	.16832	.18060
30	.19032	.21756	.24170	.27023	.28987	80	.11787	.13467	.14960	.16728	.17949
31	.18732	.21412	.23788	.26596	.28530	81	.11716	.13385	.14868	.16626	.17840
32	.18445	.21085	.23424	.26189	.28094	82	.11645	.13305	.14779	.16526	.17732
33	.18171	.20771	.23076	.25801	.27677	83	.11576	.13226	.14691	.16428	.17627
34	.17909	.20472	.22743	.25429	.27279	84	.11508	.13148	.14605	.16331	.17523
35	.17659	.20185	.22425	.25073	.26897	85	.11442	.13072	.14520	.16236	.17421
36	.17418	.19910	.22119	.24732	.26532	86	.11376	.12997	.14437	.16143	.17321
37	.17188	.19646	.21826	.24404	.26180	87	.11311	.12923	.14355	.16051	.17223
38	.16966	.19392	.21544	.24089	.25843	88	.11248	.12850	.14274	.15961	.17126
39	.16753	.19148	.21273	.23786	.25518	89	.11186	.12779	.14195	.15873	.17031
40	.16547	.18913	.21012	.23494	.25205	90	.11125	.12709	.14117	.15786	.16938
41	.16349	.18687	.20760	.23213	.24904	91	.11064	.12640	.14040	.15700	.16846
42	.16158	.18468	.20517	.22941	.24613	92	.11005	.12572	.13965	.15616	.16755
43	.15974	.18257	.20283	.22679	.24332	93	.10947	.12506	.13891	.15533	.16666
44	.15796	.18053	.20056	.22426	.24060	94	.10889	.12440	.13818	.15451	.16579
45	.15623	.17856	.19837	.22181	.23798	95	.10833	.12375	.13746	.15371	.16493
46	.15457	.17665	.19625	.21944	.23544	96	.10777	.12312	.13675	.15291	.16408
47	.15295	.17481	.19420	.21715	.23298	97	.10722	.12249	.13606	.15214	.16324
48	.15139	.17302	.19221	.21493	.23059	98	.10668	.12187	.13537	.15137	.16242
49	.14987	.17128	.19028	.21277	.22828	99	.10615	.12126	.13469	.15061	.16161
50	.14840	.16959	.18841	.21068	.22604	100	.10563	.12067	.13403	.14987	.16081

For sample sizes $n > 100$ the two-tailed critical value D_α can be found as $D_\alpha = \sqrt{-ln(\alpha/2)/2n}$.

For $\alpha = 0.05$ this is $1.358/\sqrt{n}$ and for $\alpha = 0.01$ is $1.628/\sqrt{n}$.

This table has been adapted from L. H. Miller, 'Table of percentage points of Kolmogorov statistics', *J. Amer. Statist. Assoc.*, Vol. 51 (1956), pp. 113–15, with permission of the author and publisher.

Table 21. Kolmogorov–Smirnov two-sample test

Critical values of D = maximum $|S_1(X) - S_2(X)|$, where $S_1(X)$ and $S_2(X)$ are cumulative number of observations ($\leqslant X$) in sample 1 and 2 respectively, and where sample size $n_1 = n_2 = n$. **Example:** In a two-tailed test, where $n = 27$ and $D \geqslant 10$, H_0 (that the samples are stochastically equal) may be rejected at $\alpha = 0.05$ level, i.e. a large enough D is evidence for rejection. In a one-tailed test the direction of the stochastic difference has to be predicted and the sign of D noted.

Sample size n	One-tailed test		Two-tailed test	
	0.05	0.01	0.05	0.01
3	3	∞	∞	∞
4	4	∞	4	∞
5	4	5	5	5
6	5	6	5	6
7	5	6	6	6
8	5	6	6	7
9	6	7	6	7
10	6	7	7	8
11	6	8	7	8
12	6	8	7	8
13	7	8	7	9
14	7	8	8	9
15	7	9	8	9
16	7	9	8	10
17	8	9	8	10
18	8	10	9	10
19	8	10	9	10
20	8	10	9	11
21	8	10	9	11
22	9	11	9	11
23	9	11	10	11
24	9	11	10	12
25	9	11	10	12
26	9	11	10	12
27	9	12	10	12
28	10	12	11	13
29	10	12	11	13
30	10	12	11	13

The one-tailed test has been abridged from L. A. Goodman, 'Kolmogorov–Smirnov tests for psychological research', *Psychol. Bull.*, Vol. 51 (1954), table 3, p. 167, with permission of the author and publisher.

The two-tailed test has been derived from F. J. Massey, Jr., 'The distribution of the maximum deviation between two sample cumulative step functions', *Ann. Math. Statist.*, Vol. 22 (1951), pp. 126–7, with permission of the author and publisher.

Table 22. Sign test

Probabilities that the number of signs (+ or −, whichever is smaller) will occur x or less times in a random sample of n matched pairs. H_0 is that $P(X_A > X_B) = P(X_A < X_B) = ½$, where X_A and X_B are the two 'scores' for a matched pair, and, therefore, the probabilities are based on the Binomial distribution with $p = q = ½$. **Example:** If 22 matched pairs are observed, 16 showing differences in one direction (+) and 6 in the other (−), then $P(x \leqslant 6) = 0.026$ and H_0 may be rejected at $\alpha = 0.026$ level (one-tailed test). A two-tailed test would reject H_0 at $\alpha = 0.052$ level.

Acknowledgement to Table 23

This table has been adapted from F. Wilcoxon and R. A. Wilcox, *Some rapid approximate statistical procedures*, revised 1964, table 2, p. 28, Pearl River, New York: American Cyanamid Company, with permission of the authors and publisher.

Table 22 – *continued*

n \ x	0	1	2	3	4	5	6	7	8	9	10	11	12	13	14	15
5	.031	.188	.500	.812	.969	*										
6	.016	.109	.344	.656	.891	.984	*									
7	.008	.062	.227	.500	.773	.938	.992	*								
8	.004	.035	.145	.363	.637	.855	.965	.996	*							
9	.002	.020	.090	.254	.500	.746	.910	.980	.998	*						
10	.001	.011	.055	.172	.377	.623	.828	.945	.989	.999	*					
11		.006	.033	.113	.274	.500	.726	.887	.967	.994	*	*				
12		.003	.019	.073	.194	.387	.613	.806	.927	.981	.997	*	*			
13		.002	.011	.046	.133	.291	.500	.709	.867	.954	.989	.998	*	*		
14		.001	.006	.029	.090	.212	.395	.605	.788	.910	.971	.994	.999	*	*	
15			.004	.018	.059	.151	.304	.500	.696	.849	.941	.982	.996	*	*	*
16			.002	.011	.038	.105	.227	.402	.598	.773	.895	.962	.989	.998	*	*
17			.001	.006	.025	.072	.166	.315	.500	.685	.834	.928	.975	.994	.999	*
18			.001	.004	.015	.048	.119	.240	.407	.593	.760	.881	.952	.985	.996	.999
19				.002	.010	.032	.084	.180	.324	.500	.676	.820	.916	.968	.990	.998
20				.001	.006	.021	.058	.132	.252	.412	.588	.748	.868	.942	.979	.994
21				.001	.004	.013	.039	.095	.192	.332	.500	.668	.808	.905	.961	.987
22					.002	.008	.026	.067	.143	.262	.416	.584	.738	.857	.933	.974
23					.001	.005	.017	.047	.105	.202	.339	.500	.661	.798	.895	.953
24					.001	.003	.011	.032	.076	.154	.271	.419	.581	.729	.846	.924
25						.002	.007	.022	.054	.115	.212	.345	.500	.655	.788	.885

* Values of 1.0 or approximately 1.0.

For sample sizes >25, the Normal approximation to the Binomial distribution can be used.

This table has been adapted from Helen Walker, and Joseph Lev, *Statistical Inference*, 1953, table 4, B, p. 458, New York: Holt, Rinehart and Winston, Inc., with permission of the authors and publisher.

Table 23. Wilcoxon matched-pairs signed-ranks test

Critical values of T = smaller sum of like-signed ranks (positive or negative) and their associated levels of significance. **Example:** If, say, $T = 2$ is the sum of the negative ranks when $n = 7$, H_0 (that there is no difference between the two groups) may be rejected at $\alpha = 0.05$ level of significance (two-tailed test). Similarly, H_0 may be rejected at $\alpha = 0.025$ level if it was predicted that the sum of negative ranks would be the smaller sum.

Number of matched-pairs n	Level of significance for one-tailed test, direction predicted			
	0.05	0.025	0.01	0.005
	Level of significance for two-tailed test, direction not predicted			
	0.10	0.05	0.02	0.01
5	1	–	–	–
6	2	1	–	–
7	4	2	0	–
8	6	4	2	0
9	8	6	3	2
10	11	8	5	3
11	14	11	7	5
12	17	14	10	7
13	21	17	13	10
14	26	21	16	13
15	30	25	20	16
16	36	30	24	19
17	41	35	28	23
18	47	40	33	28
19	54	46	38	32
20	60	52	43	37
21	68	59	49	43
22	75	66	56	49
23	83	73	62	55
24	92	81	69	61
25	101	90	77	68

For sample sizes $n > 25$ the sum of the ranks, T, is approximately Normally distributed with zero mean and unit variance.

Table 24. Mann–Whitney U-test (probabilities)

Probabilities that U is less or equal to an observed value of U when H_0 states that two samples come from populations which are stochastically equal. U = number of times a 'score' in the sample of size m precedes a 'score' in the sample of size n ($n \geqslant m$) when all the $n + m$ 'scores' are jointly ranked. **Example:** If the 'scores' in the sample with $m = 2$ observations are: 6 and 7 and the 'scores' in the sample with $n = 3$ observations are: 3, 5 and 9, then the joint ranking is: 3, 5, 6, 7 and 9 and $U = 0 + 0 + 2 = 2$ as there are no 'scores' from the first sample preceding scores 3 and 5 from the second sample, and two scores, i.e. 6 and 7, preceding 9. $P(U \leqslant 2) = 0.400$, and, therefore, H_0 cannot be rejected at $\alpha = 0.05$ level (one-tailed test). For a two-tailed test the probability has to be doubled. If U is larger than the values given in the table, change the direction of the one-tailed hypothesis, reverse the order in which the preceding 'scores' are counted and thus obtain $U' = nm - U$. Then test U' as above.

$n = 3$

U \ m	1	2	3
0	.250	.100	.050
1	.500	.200	.100
2	.750	.400	.200
3		.600	.350
4			.500
5			.650

$n = 4$

U \ m	1	2	3	4
0	.200	.067	.028	.014
1	.400	.133	.057	.029
2	.600	.267	.114	.057
3		.400	.200	.100
4		.600	.314	.171
5			.429	.243
6			.571	.343
7				.443
8				.557

$n = 5$

U \ m	1	2	3	4	5
0	.167	.047	.018	.008	.004
1	.333	.095	.036	.016	.008
2	.500	.190	.071	.032	.016
3	.667	.286	.125	.056	.028
4		.429	.196	.095	.048
5		.571	.286	.143	.075
6			.393	.206	.111
7			.500	.278	.155
8			.607	.365	.210
9				.452	.274
10				.548	.345
11					.421
12					.500
13					.579

$n = 6$

U \ m	1	2	3	4	5	6
0	.143	.036	.012	.005	.002	.001
1	.286	.071	.024	.010	.004	.002
2	.428	.143	.048	.019	.009	.004
3	.571	.214	.083	.033	.015	.008
4		.321	.131	.057	.026	.013
5		.429	.190	.086	.041	.021
6		.571	.274	.129	.063	.032
7			.357	.176	.089	.047
8			.452	.238	.123	.066
9			.548	.305	.165	.090
10				.381	.214	.120
11				.457	.268	.155
12				.545	.331	.197
13					.396	.242
14					.465	.294
15					.535	.350
16						.409
17						.469
18						.531

This table has been reproduced from H. B. Mann and D. R. Whitney, 'On a test of whether one of two random variables is stochastically larger than the other', *Ann. Math. Statist.*, Vol. 18 (1947), pp. 52–4, with permission of the authors and publisher.

Table 24 – *continued*

$n = 7$

U \ m	1	2	3	4	5	6	7
0	.125	.028	.008	.003	.001	.001	.000
1	.250	.056	.017	.006	.003	.001	.001
2	.375	.111	.033	.012	.005	.002	.001
3	.500	.167	.058	.021	.009	.004	.002
4	.625	.250	.092	.036	.015	.007	.003
5		.333	.133	.055	.024	.011	.006
6		.444	.192	.082	.037	.017	.009
7		.556	.258	.115	.053	.026	.013
8			.333	.158	.074	.037	.019
9			.417	.206	.101	.051	.027
10			.500	.264	.134	.069	.036
11			.583	.324	.172	.090	.049
12				.394	.216	.117	.064
13				.464	.265	.147	.082
14				.538	.319	.183	.104
15					.378	.223	.130
16					.438	.267	.159
17					.500	.314	.191
18					.562	.365	.228
19						.418	.267
20						.473	.310
21						.527	.355
22							.402
23							.451
24							.500
25							.549

$n = 8$

U \ m	1	2	3	4	5	6	7	8	Normal z	Normal α
0	.111	.022	.006	.002	.001	.000	.000	.000	3.308	.001
1	.222	.044	.012	.004	.002	.001	.000	.000	3.203	.001
2	.333	.089	.024	.008	.003	.001	.001	.000	3.098	.001
3	.444	.133	.042	.014	.005	.002	.001	.001	2.993	.001
4	.556	.200	.067	.024	.009	.004	.002	.001	2.888	.002
5		.267	.097	.036	.015	.006	.003	.001	2.783	.003
6		.356	.139	.055	.023	.010	.005	.002	2.678	.004
7		.444	.188	.077	.033	.015	.007	.003	2.573	.005
8		.556	.248	.107	.047	.021	.010	.005	2.468	.007
9			.315	.141	.064	.030	.014	.007	2.363	.009
10			.387	.184	.085	.041	.020	.010	2.258	.012
11			.461	.230	.111	.054	.027	.014	2.153	.016
12			.539	.285	.142	.071	.036	.019	2.048	.020
13				.341	.177	.091	.047	.025	1.943	.026
14				.404	.217	.114	.060	.032	1.838	.033
15				.467	.262	.141	.076	.041	1.733	.041
16				.533	.311	.172	.095	.052	1.628	.052
17					.362	.207	.116	.065	1.523	.064
18					.416	.245	.140	.080	1.418	.078
19					.472	.286	.168	.097	1.313	.094
20					.528	.331	.198	.117	1.208	.113
21						.377	.232	.139	1.102	.135
22						.426	.268	.164	0.998	.159
23						.475	.306	.191	0.893	.185
24						.525	.347	.221	0.788	.215
25							.389	.253	0.683	.247
26							.433	.287	0.578	.282
27							.478	.323	0.473	.318
28							.522	.360	0.368	.356
29								.399	0.263	.396
30								.439	0.158	.437
31								.480	0.052	.481
32								.520		

Table 25. Mann–Whitney U-test (critical values)

Critical values of U for $m \leqslant 20$ and $9 \leqslant n \leqslant 20$ and their associated levels of significance (also see notes to Table 24). **Example:** If, say, $m = 4$ and $n = 14$, H_0 may be rejected at $\alpha = 0.05$ level in a one-tailed test, and at $\alpha = 0.10$ level in a two-tailed test, when observed $U \leqslant 11$.

One-tailed test (0.05) with direction predicted
Two-tailed test (0.10) with direction not predicted

m \ n	9	10	11	12	13	14	15	16	17	18	19	20
2	1	1	1	2	2	2	3	3	3	4	4	4
3	3	4	5	5	6	7	7	8	9	9	10	11
4	6	7	8	9	10	11	12	14	15	16	17	18
5	9	11	12	13	15	16	18	19	20	22	23	25
6	12	14	16	17	19	21	23	25	26	28	30	32
7	15	17	19	21	24	26	28	30	33	35	37	39
8	18	20	23	26	28	31	33	36	39	41	44	47
9	21	24	27	30	33	36	39	42	45	48	51	54
10	24	27	31	34	37	41	44	48	51	55	58	62
11	27	31	34	38	42	46	50	54	57	61	65	69
12	30	34	38	42	47	51	55	60	64	68	72	77
13	33	37	42	47	51	56	61	65	70	75	80	84
14	36	41	46	51	56	61	66	71	77	82	87	92
15	39	44	50	55	61	66	72	77	83	88	94	100
16	42	48	54	60	65	71	77	83	89	95	101	107
17	45	51	57	64	70	77	83	89	96	102	109	115
18	48	55	61	68	75	82	88	95	102	109	116	123
19	51	58	65	72	80	87	94	101	109	116	123	130
20	54	62	69	77	84	92	100	107	115	123	130	138

One-tailed test (0.025) with direction predicted
Two-tailed test (0.05) with direction not predicted

m \ n	9	10	11	12	13	14	15	16	17	18	19	20
2	0	0	0	1	1	1	1	1	2	2	2	2
3	2	3	3	4	4	5	5	6	6	7	7	8
4	4	5	6	7	8	9	10	11	11	12	13	13
5	7	8	9	11	12	13	14	15	17	18	19	20
6	10	11	13	14	16	17	19	21	22	24	25	27
7	12	14	16	18	20	22	24	26	28	30	32	34
8	15	17	19	22	24	26	29	31	34	36	38	41
9	17	20	23	26	28	31	34	37	39	42	45	48
10	20	23	26	29	33	36	39	42	45	48	52	55
11	23	26	30	33	37	40	44	47	51	55	58	62
12	26	29	33	37	41	45	49	53	57	61	65	69
13	28	33	37	41	45	50	54	59	63	67	72	76
14	31	36	40	45	50	55	59	64	67	74	78	83
15	34	39	44	49	54	59	64	70	75	80	85	90
16	37	42	47	53	59	64	70	75	81	86	92	98
17	39	45	51	57	63	67	75	81	87	93	99	105
18	42	48	55	61	67	74	80	86	93	99	106	112
19	45	52	58	65	72	78	85	92	99	106	113	119
20	48	55	62	69	76	83	90	98	105	112	119	127

Table 25 – *continued*

One-tailed test (0.01) with direction predicted
Two-tailed test (0.02) with direction not predicted

m \ n	9	10	11	12	13	14	15	16	17	18	19	20
2					0	0	0	0	0	0	1	1
3	1	1	1	2	2	2	3	3	4	4	4	5
4	3	3	4	5	5	6	7	7	8	9	9	10
5	5	6	7	8	9	10	11	12	13	14	15	16
6	7	8	9	11	12	13	15	16	18	19	20	22
7	9	11	12	14	16	17	19	21	23	24	26	28
8	11	13	15	17	20	22	24	26	28	30	32	34
9	14	16	18	21	23	26	28	31	33	36	38	40
10	16	19	22	24	27	30	33	36	38	41	44	47
11	18	22	25	28	31	34	37	41	44	47	50	53
12	21	24	28	31	35	38	42	46	49	53	56	60
13	23	27	31	35	39	43	47	51	55	59	63	67
14	26	30	34	38	43	47	51	56	60	65	69	73
15	28	33	37	42	47	51	56	61	66	70	75	80
16	31	36	41	46	51	56	61	66	71	76	82	87
17	33	38	44	49	55	60	66	71	77	82	88	93
18	36	41	47	53	59	65	70	76	82	88	94	100
19	38	44	50	56	63	69	75	82	88	94	101	107
20	40	47	53	60	67	73	80	87	93	100	107	114

One-tailed test (0.001) with direction predicted
Two-tailed test (0.002) with direction not predicted

m \ n	9	10	11	12	13	14	15	16	17	18	19	20
4		0	0	0	1	1	1	2	2	3	3	3
5	1	1	2	2	3	3	4	5	5	6	7	7
6	2	3	4	4	5	6	7	8	9	10	11	12
7	3	5	6	7	8	9	10	11	13	14	15	16
8	5	6	8	9	11	12	14	15	17	18	20	21
9	7	8	10	12	14	15	17	19	21	23	25	26
10	8	10	12	14	17	19	21	23	25	27	29	32
11	10	12	15	17	20	22	24	27	29	32	34	37
12	12	14	17	20	23	25	28	31	34	37	40	42
13	14	17	20	23	26	29	32	35	38	42	45	48
14	15	19	22	25	29	32	36	39	43	46	50	54
15	17	21	24	28	32	36	40	43	47	51	55	59
16	19	23	27	31	35	39	43	48	52	56	60	65
17	21	25	29	34	38	43	47	52	57	61	66	70
18	23	27	32	37	42	46	51	56	61	66	71	76
19	25	29	34	40	45	50	55	60	66	71	77	82
20	26	32	37	42	48	54	59	65	70	76	82	88

These tables have been adapted and abridged from D. Aube, 'Extended tables for the Mann–Whitney Statistic', *Bulletin of the Institute of Educational Research*, Indiana University, Vol. 1, No. 2 (1953), tables 1, 3, 5 and 7, with the permission of the author and publisher.

Table 26. Kruskal–Wallis one-way analysis of variance by ranks

Probabilities that $H \geqslant$ an observed value of H, based on H_0 stating that all samples come from the same or identical populations with respect to averages. H is defined as:

$$H = \frac{12}{n(n+1)}\left[\sum_{i=1}^{k} \frac{R_i^2}{n_i}\right] - 3(n+1)$$

where k = number of samples, n_i = number of observations in ith sample, $n = \Sigma n_i$ = number of observations in all samples combined and R_i = sum of ranks in ith sample (column). All n observations are ranked jointly, the smallest is given rank 1, and the largest rank n. **Example**: If, say, $H \geqslant 6.7455$ when $n_1 = 4$, $n_2 = 3$ and $n_3 = 3$, H_0 may be rejected at $\alpha = 0.10$ level (one-tailed test).

Sample sizes n_1	n_2	n_3	H	P
2	1	1	2.7000	.500
2	2	1	3.6000	.200
2	2	2	4.5714	.067
			3.7143	.200
3	1	1	3.2000	.300
3	2	1	4.2857	.100
			3.8571	.133
3	2	2	5.3572	.029
			4.7143	.048
			4.5000	.067
			4.4643	.105
3	3	1	5.1429	.043
			4.5714	.100
			4.0000	.129
3	3	2	6.2500	.011
			5.3611	.032
			5.1389	.061
			4.5556	.100
			4.2500	.121
3	3	3	7.2000	.004
			6.4889	.011
			5.6889	.029
			5.6000	.050
			5.0667	.086
			4.6222	.100
4	1	1	3.5714	.200
4	2	1	4.8214	.057
			4.5000	.076
			4.0179	.114
4	2	2	6.0000	.014
			5.3333	.033
			5.1250	.052
			4.4583	.100
			4.1667	.105
4	3	1	5.8333	.021
			5.2083	.050
			5.0000	.057
			4.0556	.093
			3.8889	.129
4	3	2	6.4444	.008
			6.3000	.011
			5.4444	.046
			5.4000	.051
			4.5111	.098
			4.4444	.102
4	3	3	6.7455	.010
			6.7091	.013
			5.7909	.046
			5.7273	.050
			4.7091	.092
			4.7000	.101
4	4	1	6.6667	.010
			6.1667	.022
			4.9667	.048
			4.8667	.054
			4.1667	.082
			4.0667	.102
4	4	2	7.0364	.006
			6.8727	.011
			5.4545	.046
			5.2364	.052
			4.5545	.098
			4.4455	.103
4	4	3	7.1439	.010
			7.1364	.011
			5.5985	.049
			5.5758	.051
			4.5455	.099
			4.4773	.102
4	4	4	7.6538	.008
			7.5385	.011
			5.6923	.049
			5.6538	.054
			4.6539	.097
			4.5001	.104
5	1	1	3.8571	.143
5	2	1	5.2500	.036
			5.0000	.048
			4.4500	.071
			4.2000	.095
			4.0500	.119
5	2	2	6.5333	.008
			6.1333	.013
			5.1600	.034
			5.0400	.056
			4.3733	.090
			4.2933	.122
5	3	1	6.4000	.012
			4.9600	.048
			4.8711	.052
			4.0178	.095
			3.8400	.123
5	3	2	6.9091	.009
			6.8218	.010
			5.2509	.049
			5.1055	.052
			4.6509	.091
			4.4945	.101
5	3	3	7.0788	.009
			6.9818	.011
			5.6485	.049
			5.5152	.051
			4.5333	.097
			4.4121	.109
5	4	1	6.9545	.008
			6.8400	.011
			4.9855	.044
			4.8600	.056
			3.9873	.098
			3.9600	.102
5	4	2	7.2045	.009
			7.1182	.010
			5.2727	.049
			5.2682	.050
			4.5409	.098
			4.5182	.101
5	4	3	7.4449	.010
			7.3949	.011
			5.6564	.049
			5.6308	.050
			4.5487	.099
			4.5231	.103
5	4	4	7.7604	.009
			7.7440	.011
			5.6571	.049
			5.6176	.050
			4.6187	.100
			4.5527	.102
5	5	1	7.3091	.009
			6.8364	.011
			5.1273	.046
			4.9091	.053
			4.1091	.086
			4.0364	.105
5	5	2	7.3385	.010
			7.2692	.010
			5.3385	.047
			5.2462	.051
			4.6231	.097
			4.5077	.100
5	5	3	7.5780	.010
			7.5429	.010
			5.7055	.046
			5.6264	.051
			4.5451	.100
			4.5363	.102
5	5	4	7.8229	.010
			7.7914	.010
			5.6657	.049
			5.6429	.050
			4.5229	.099
			4.5200	.101
5	5	5	8.0000	.009
			7.9800	.010
			5.7800	.049
			5.6600	.051
			4.5600	.100
			4.5000	.102

When $k > 3$ and $n_i > 5$ for all i, H is distributed like χ^2 with $\nu = k - 1$ degrees of freedom and, therefore, probabilities associated with observed values of H may be found in Table 13.

This table has been adapted and abridged from W. H. Kruskal and W. A. Wallis, 'Use of ranks in one-criterion variance analysis', *J. Amer. Statist. Assoc.*, Vol. 47 (1952), table 6.1, pp. 614–17, with permission of the authors and publisher. The table contains the corrections given later by the authors in *J. Amer. Statist. Assoc.*, Vol. 48 (1953), p. 910.

(d) Random numbers

Table 27. Random digits

Random digits 0 to 9, which are 'blocked' for convenience, may be used in any systematic way; e.g. if a random sample of 5 is required from population consisting of 83 members, the first two columns may be used to identify the members of the population, i.e. numbers 01 to 83, and the selected numbers are: 29, 12, 02, 69 and 11. Numbers greater than 83 may be ignored. When the first two columns are exhausted, columns 3 and 4 may be used, etc.

29	32	95	99	57	98	08	36	97	08	65	30	47	22	00	38	60	10	01	10
12	11	80	16	17	01	03	97	59	73	74	98	73	65	85	59	74	66	37	58
87	58	22	25	55	35	72	79	28	15	69	17	42	98	72	05	47	12	40	99
02	92	42	87	57	53	53	34	55	75	83	64	09	10	19	33	29	57	62	98
69	28	63	73	98	45	61	10	43	20	08	10	43	16	81	17	62	99	09	16
11	95	68	77	86	91	76	11	63	34	15	08	35	39	37	12	74	15	00	10
06	43	41	02	13	65	23	94	48	88	88	87	03	90	77	68	98	09	17	22
68	55	98	08	39	59	85	46	66	13	42	90	86	13	29	12	38	48	27	54
41	01	06	65	10	29	29	91	86	24	45	59	04	88	17	68	31	01	91	13
46	75	71	76	88	04	42	94	41	42	39	79	14	46	13	49	37	18	28	08
80	14	13	43	24	47	61	47	42	24	24	82	12	23	54	81	33	18	96	89
30	56	60	77	80	33	67	68	31	67	73	23	45	30	55	81	51	87	68	58
53	50	41	02	98	49	97	32	43	55	75	33	51	20	99	64	76	20	80	98
84	14	75	87	37	58	51	94	06	73	27	94	23	76	77	81	72	90	45	41
08	27	89	33	87	52	24	57	50	22	22	76	60	05	79	86	58	83	88	41
97	08	50	16	41	67	40	56	13	12	68	67	36	22	08	55	76	86	45	67
97	08	37	42	48	95	90	48	34	88	19	66	38	94	64	95	07	78	23	86
70	15	04	10	34	95	57	63	75	82	88	74	28	24	66	99	52	65	36	98
06	38	31	17	38	24	98	52	67	04	95	54	89	79	45	28	05	18	60	17
63	87	79	25	86	56	74	17	45	32	53	62	09	04	86	65	87	48	82	02
17	00	56	31	14	18	56	97	91	78	85	82	06	24	88	49	17	68	51	50
17	76	35	38	19	24	47	21	09	43	09	72	02	64	66	06	78	21	70	41
57	77	32	13	60	37	68	66	11	23	30	62	97	71	02	20	13	22	00	40
35	86	97	84	91	77	73	03	37	77	50	24	54	51	40	20	66	16	34	84
72	68	64	77	89	72	77	67	45	72	25	56	78	69	72	63	86	52	07	43
91	01	78	50	50	91	99	15	36	02	74	42	55	33	19	88	35	17	58	37
70	37	55	94	53	05	78	53	23	29	15	57	70	30	88	63	20	12	64	38
11	06	17	48	24	57	50	76	81	77	30	12	92	27	19	32	63	70	97	80
60	37	89	98	61	05	51	89	47	28	34	83	98	44	66	96	84	64	64	92
37	41	11	09	04	84	38	51	91	49	23	78	53	95	40	17	73	23	04	70
28	97	38	27	97	54	95	94	54	79	93	88	00	82	39	61	93	78	07	88
14	29	17	18	84	03	10	62	15	70	01	15	06	30	97	79	55	98	79	39
81	70	53	83	20	25	26	56	55	56	33	58	74	21	76	94	24	80	12	50
08	20	90	25	43	22	81	74	51	76	53	39	59	35	34	46	55	54	73	50
61	95	25	85	66	34	76	39	98	88	45	57	64	11	17	06	43	35	27	09
64	58	31	05	45	77	25	20	02	09	36	87	63	01	10	08	01	19	19	06
75	49	97	87	79	31	66	57	89	56	56	97	71	43	65	62	36	77	50	87
66	95	10	78	42	24	91	82	74	29	00	53	44	70	18	23	48	09	90	99
85	37	61	48	07	99	13	01	16	94	37	31	28	96	59	77	62	24	95	84
06	87	15	09	48	31	18	66	87	11	19	71	67	20	93	92	02	96	15	65
11	15	95	59	69	81	75	75	88	69	95	12	75	69	18	10	60	35	31	47
03	64	44	33	46	16	02	28	14	33	61	57	28	33	96	47	49	86	85	83
68	89	57	51	94	84	09	80	37	90	52	99	85	52	49	66	63	69	11	31
43	13	09	12	00	65	69	54	11	00	20	94	22	93	90	16	82	64	27	46
42	68	71	56	74	17	71	63	80	81	02	41	49	27	92	44	44	13	45	21
12	55	09	80	30	50	34	96	31	71	19	21	79	42	17	57	04	04	19	00
88	84	87	74	01	39	99	02	75	76	61	88	97	89	06	97	15	70	26	27
49	27	92	08	87	65	12	32	27	96	11	26	30	88	48	89	29	73	50	47
46	51	54	92	06	44	85	83	14	78	68	83	33	17	03	10	99	10	17	34
34	96	78	90	18	41	44	69	10	30	48	98	32	76	12	81	29	83	02	87

Table 27 – *continued*

80	07	15	41	15	37	42	39	24	45	48	73	61	15	44	74	40	27	26
39	08	51	67	63	03	76	76	86	09	39	32	62	77	60	85	37	14	69
51	32	57	06	49	13	01	25	98	83	44	96	92	78	37	24	49	35	54
84	46	17	46	71	53	88	78	30	71	53	85	55	10	93	40	05	66	72
04	88	20	78	89	94	31	36	83	74	51	25	28	43	54	76	57	08	21
21	45	86	26	12	21	28	37	56	47	86	18	38	39	18	89	99	62	81
71	38	27	31	40	52	36	03	51	54	83	14	51	17	86	77	66	84	50
78	50	39	32	55	17	25	06	90	90	69	48	70	68	22	07	85	07	95
22	76	93	40	26	30	77	61	71	74	81	13	73	21	99	00	47	52	43
25	21	70	62	69	05	05	58	75	92	85	60	50	87	81	35	80	83	42
96	79	06	87	51	04	17	61	42	12	64	77	45	06	55	68	19	39	17
97	76	01	89	33	70	46	23	44	83	99	55	95	03	41	89	33	49	89
78	03	18	58	00	47	18	01	33	49	99	55	54	70	65	34	76	58	86
09	63	31	80	30	17	11	75	34	81	25	45	91	80	50	25	64	70	05
61	33	89	72	78	98	26	56	88	66	51	69	71	48	13	71	40	57	31
64	83	61	76	37	68	22	25	09	82	53	59	78	66	81	66	45	56	64
18	93	65	67	39	81	96	44	68	46	96	50	08	71	70	81	23	32	89
86	84	70	40	22	89	25	42	62	69	95	98	59	26	69	55	33	62	91
96	57	56	48	81	92	77	95	43	50	29	89	07	58	10	83	66	04	15
54	35	65	28	09	99	04	41	86	60	69	54	82	74	49	86	82	25	07
18	79	09	01	55	60	31	19	19	48	01	89	54	63	96	70	99	15	71
19	78	77	63	36	52	38	88	16	92	23	42	49	79	27	15	09	94	49
55	71	79	75	30	29	13	32	60	07	33	73	61	89	63	64	17	15	21
38	58	83	62	94	73	84	48	95	17	79	74	78	38	09	37	35	75	74
78	29	66	85	65	45	79	70	88	92	73	24	71	71	63	70	47	56	70
87	55	81	22	04	62	21	45	81	82	43	96	17	70	61	80	59	10	59
06	98	70	24	03	20	67	45	67	65	04	61	76	89	25	13	73	06	41
33	08	62	21	90	70	72	16	01	23	26	05	10	33	23	23	03	07	46
54	03	25	45	50	40	58	15	41	07	16	24	16	63	46	64	27	85	27
68	90	88	08	25	70	23	82	53	40	51	91	84	67	84	08	09	76	19
90	18	00	18	76	88	55	07	52	00	30	04	83	72	04	74	87	56	90
70	07	33	78	52	59	92	46	58	33	61	42	31	47	58	89	32	02	55
19	13	05	69	12	74	49	85	21	49	18	11	60	96	94	04	74	26	23
95	70	86	00	19	44	74	51	22	34	63	14	11	30	48	54	71	78	97
65	12	41	20	32	33	72	70	71	24	51	39	43	28	90	51	14	46	17
15	53	57	75	61	54	95	63	75	51	28	43	39	55	90	58	01	50	31
60	27	72	94	00	25	71	09	76	19	66	69	44	09	39	12	60	43	02
57	91	58	68	24	78	33	54	25	46	08	87	72	85	28	98	89	67	68
40	15	42	80	71	35	81	75	95	40	04	85	70	88	19	44	75	50	63
23	97	89	48	74	96	60	10	40	24	33	88	86	93	30	79	96	32	25
48	25	55	19	87	97	39	79	66	73	50	78	72	75	08	78	66	69	13
24	58	57	51	61	90	39	52	91	33	77	67	76	78	40	42	05	70	73
60	22	38	11	98	95	66	00	95	19	32	99	90	77	55	50	86	94	41
84	89	06	96	10	47	83	22	11	81	19	13	48	21	71	99	16	81	88
30	80	70	60	93	09	74	04	99	72	67	91	91	75	20	36	08	45	28
23	95	78	32	20	71	90	24	20	66	09	27	14	97	94	78	67	45	20
48	52	58	73	69	63	54	77	76	89	09	15	50	05	85	91	12	10	12
33	69	72	87	15	96	24	09	14	84	41	57	16	17	78	18	46	46	23
71	71	53	72	84	65	86	16	70	43	62	10	33	15	61	60	80	73	18
29	53	27	21	49	53	31	68	21	10	17	47	35	74	84	18	58	07	17
17	70	60	84	24	50	82	33	67	40	15	88	50	22	54	28	39	46	14
98	37	60	93	52	27	20	93	10	62	90	69	27	96	44	54	01	13	81
16	39	86	14	17	56	74	44	76	20	77	74	52	52	56	06	99	78	52
53	17	93	61	99	15	08	47	04	09	46	95	53	02	57	60	02	02	99
05	38	06	80	55	75	49	12	95	96	98	63	46	51	49	74	97	71	95

Table 28. Random standardised Normal deviates

Random sample of numbers from Normal distribution with $\mu = 0$ and $\sigma = 1$ which can be used in any systematic way; e.g. to obtain a random sample of five, the first five numbers in the first row may be taken, i.e. −0.262, 1.560, 1.366, −0.222 and −0.063. If a random sample of five is required from a Normal distribution with $\mu = 10$ and $\sigma = 2$, say, the five numbers selected above (z) may be used again but must be transformed using the relationship $x = \sigma z + \mu$, i.e. $x_1 = 2(-0.262) + 10 = 9.476$, etc.

−0.262	1.560	1.366	−0.222	−0.063	−0.225	−0.094	0.280	−1.057	0.322
0.233	0.279	−2.806	1.493	−0.911	−1.002	−0.345	0.075	−1.472	1.760
−0.476	−0.806	0.546	1.428	−1.234	0.225	1.571	0.483	1.554	2.167
1.895	−0.559	1.690	0.791	0.142	0.568	−1.460	−1.555	−1.843	−0.544
2.353	1.590	2.182	−0.801	−0.217	0.066	−0.071	1.079	0.426	−0.517
0.028	−0.756	−0.170	0.171	0.617	−0.268	−0.390	−0.811	−0.183	−0.441
−0.619	−0.186	−0.971	0.559	1.775	−0.002	−0.621	−1.517	−0.410	−1.008
−0.779	−1.248	0.742	0.177	−0.232	0.702	−1.593	0.538	−0.007	−2.045
−0.269	0.856	−2.078	0.472	1.313	0.044	−0.369	0.183	0.733	−0.229
2.123	0.628	0.434	−0.018	1.271	0.174	−0.633	−0.012	−1.147	0.125
0.658	−0.899	0.822	−0.605	0.276	−1.425	−0.334	−1.032	−0.028	−0.787
0.394	−0.396	0.063	−0.990	−0.991	2.256	0.159	1.984	0.659	1.784
−0.302	0.944	−1.335	0.989	0.752	−0.510	0.442	−0.577	−0.114	0.129
−0.593	−1.112	0.616	0.696	0.780	1.682	1.021	−0.791	0.934	1.304
−1.655	1.664	1.102	1.585	1.088	−0.499	−1.577	−0.915	0.350	0.390
−1.014	0.980	0.077	0.945	1.016	1.848	2.643	0.963	0.670	1.678
1.721	1.166	−0.182	−0.259	−0.783	0.346	−0.741	0.023	−0.473	−0.161
0.201	0.207	−0.702	−0.236	1.100	1.460	0.776	−0.636	0.215	−1.502
−0.159	1.047	0.343	−0.752	−0.629	−0.668	−1.392	0.264	−0.810	0.836
−2.066	−1.648	1.103	−0.628	1.072	0.600	2.679	−0.583	1.704	−0.994
−0.439	−0.888	1.373	−0.802	0.041	−0.277	−0.904	0.760	−0.487	0.275
−0.674	−3.099	−1.221	−0.351	1.821	0.127	0.666	0.999	−0.016	0.452
−0.539	0.684	−0.128	−1.037	−0.276	0.196	−2.021	0.743	−0.818	1.134
−0.007	1.137	−0.089	−0.562	−1.008	−0.637	−0.901	−0.316	1.884	−0.846
1.274	−0.590	−0.335	1.555	−1.010	0.185	0.966	0.030	−0.244	−0.174
0.954	0.859	1.531	1.154	−2.578	−1.639	−0.249	−0.815	0.040	2.088
−0.588	1.693	−0.642	1.610	0.376	0.231	−0.482	−0.388	0.304	−1.578
−1.769	−0.381	−0.856	0.230	−0.159	0.322	−0.598	2.611	−1.599	−1.094
−1.186	−0.555	−0.319	−0.752	0.667	0.362	−0.611	−0.535	−0.604	0.252
1.354	−2.295	1.162	−2.210	0.126	−0.371	0.931	0.871	1.226	0.294
−0.690	−1.919	−2.066	0.715	−0.059	−1.536	−0.332	1.419	−1.477	−0.360
1.763	0.378	0.317	−0.589	−0.809	−1.665	0.709	2.375	0.081	1.800
0.292	0.595	0.963	−1.686	1.369	−1.379	0.789	−1.038	0.437	−1.615
0.407	0.281	−0.515	−0.002	1.787	−2.283	0.022	0.120	−0.741	0.053
−0.635	0.743	−0.141	0.173	−0.341	2.323	−1.863	−0.566	1.615	−0.210
0.694	1.178	−1.020	−0.531	0.219	0.932	−0.734	0.078	0.300	0.446
1.161	−0.212	0.211	0.347	−0.422	−0.280	1.811	−0.952	−0.308	1.569
2.075	−0.211	−1.026	1.136	−0.751	0.937	−1.560	−0.163	0.186	−1.156
1.046	−0.495	0.405	0.965	0.902	0.775	2.862	−0.179	−0.174	0.095
−0.111	−1.227	−0.475	0.988	0.747	−2.243	−0.481	0.278	−1.931	0.394
−0.760	0.440	1.332	0.660	1.621	2.075	0.066	0.085	1.548	−1.032
0.120	−0.559	0.462	−1.810	−0.172	−0.803	−0.489	0.111	−1.513	−0.326
−0.304	1.423	−0.403	0.560	0.921	−1.588	−1.194	−0.452	1.990	−0.710
−0.710	−0.136	1.192	−1.486	0.103	−1.881	0.837	−1.327	0.820	−1.882
1.913	−1.646	1.853	−1.737	0.356	−0.150	0.456	1.117	−1.028	−0.603
−0.012	−1.196	−0.117	0.798	0.574	−1.269	−0.036	1.327	−0.174	0.243
−1.614	0.448	−1.686	0.537	−0.047	1.993	0.485	1.099	−1.149	0.145
−1.058	0.498	0.226	0.863	−1.073	−1.148	1.422	−0.020	−0.020	−0.069
0.628	1.188	−0.652	−0.960	0.159	−0.795	0.623	1.106	0.863	−0.845
0.069	0.551	−0.610	1.477	−0.650	−0.253	−1.326	0.299	1.396	−0.321
−1.113	0.489	0.359	1.018	0.551	−0.529	0.071	−0.755	0.493	−0.959
0.669	−0.248	1.252	−0.396	0.076	1.403	0.058	0.946	−0.711	−0.189
−1.086	−0.008	−0.691	1.223	−0.589	0.955	0.508	0.676	1.513	−0.261
0.363	0.095	−0.841	−0.772	−0.315	0.823	1.514	0.267	0.494	−0.234
0.164	−0.905	0.669	0.687	0.837	0.199	−0.298	−2.013	−1.271	0.533
−0.604	−1.724	0.120	0.200	−0.303	1.536	−0.519	−0.478	−0.530	−0.812
−0.012	0.467	0.435	0.763	0.184	0.001	0.023	−0.182	−1.210	1.170
0.523	1.156	−0.010	0.639	−1.450	−1.101	0.071	0.520	0.923	1.744
0.195	1.624	0.316	−0.604	−0.573	0.282	0.797	−1.284	−2.068	−0.335
0.676	−0.328	−0.440	−0.831	0.826	0.507	−0.402	−0.560	0.333	1.137

Table 28 – *continued*

−1.227	−1.421	0.928	0.916	−0.930	0.159	0.530	1.137	−1.005	−0.14
1.072	−1.443	1.851	0.296	−0.805	0.474	−0.986	0.110	−1.281	−0.58
1.725	−0.076	−1.947	0.935	−0.222	0.027	−1.751	2.201	−1.416	−0.45
1.234	0.144	−1.104	0.620	−0.028	−0.080	0.062	0.292	−0.806	1.06
0.073	−1.481	−0.341	0.315	−0.720	1.686	−0.816	−0.425	−1.257	0.50
0.091	−0.239	0.166	−0.860	0.131	0.687	−0.944	−1.585	0.487	−3.24
0.171	1.816	0.084	−1.476	−0.028	−0.073	−1.065	−0.285	0.771	−0.51
−0.366	−0.630	0.515	0.147	−0.445	0.292	0.477	−0.690	−1.066	1.35
0.747	−1.466	−0.916	−0.313	−1.638	−1.427	−1.432	0.477	1.378	0.69
−2.076	0.055	−1.327	0.613	0.464	−1.589	−0.396	0.152	1.160	−0.04
−0.909	−0.240	0.207	−2.380	−0.127	−0.247	0.073	−1.067	−0.303	0.55
−0.191	0.692	−0.049	−1.281	0.843	−0.516	−0.473	−0.064	−0.058	0.69
−0.637	−1.506	−0.433	−0.252	−1.091	−0.393	0.098	−0.814	−1.952	−0.37
−0.585	−1.215	1.496	0.215	0.870	−0.677	0.887	−0.219	−0.808	2.165
1.336	−2.144	−0.332	−0.421	−0.649	−0.911	−0.593	−0.374	−0.555	−0.30
−0.483	−1.161	−0.914	0.188	−1.778	0.482	0.195	2.165	−0.681	−0.854
1.436	0.492	0.126	−1.793	−1.303	1.691	−0.844	−0.705	−0.262	−0.084
0.075	0.051	−0.689	0.440	−0.013	−0.284	−2.182	−0.425	−0.762	0.33
1.515	−0.790	−0.313	−0.342	−1.757	0.420	−0.548	−0.458	−0.232	−1.399
−0.926	1.020	−0.907	1.676	0.742	−1.523	−1.138	0.329	1.341	0.880
−0.791	2.030	0.953	−0.704	−0.030	2.294	1.010	−0.125	−0.293	−0.794
−1.011	−1.066	−2.273	0.306	−1.055	−0.233	−0.686	0.006	1.468	−0.465
1.788	−0.861	0.449	0.898	−1.471	−0.185	−0.584	0.785	−0.759	−0.569
1.517	1.092	1.095	−1.359	−0.961	1.239	−1.118	−1.988	−1.816	1.096
0.229	0.870	−1.943	−0.033	−0.709	−1.185	−0.442	−0.420	−0.334	−0.04
0.068	−0.584	0.279	0.149	−0.681	−0.021	0.460	−0.524	1.506	−1.442
0.003	−0.741	−0.215	−0.234	−0.819	−0.620	−0.005	−0.884	0.780	0.612
−0.026	2.155	−0.669	0.681	−0.176	−0.706	0.567	0.243	−0.887	−0.647
−0.181	−0.220	0.441	0.782	0.034	−0.798	−0.178	−0.909	−0.262	−0.465
−1.118	0.812	−0.161	−0.166	−0.485	0.632	0.101	−1.754	−1.791	0.463
−1.694	−0.563	0.298	−0.270	1.264	−1.792	−0.485	−0.933	0.308	0.650
0.787	0.016	1.012	0.419	0.180	3.001	1.297	−0.472	0.558	0.495
0.968	0.007	1.136	−0.259	2.315	0.839	0.022	2.469	0.775	0.266
−0.840	−0.713	−0.514	−1.634	−1.926	−0.435	0.598	1.240	1.088	0.298
0.716	1.153	0.638	0.391	0.742	−1.456	0.288	−0.634	−0.941	0.098
−0.387	0.894	−0.525	0.790	1.254	0.943	0.194	−0.421	0.115	0.683
0.405	2.851	0.354	0.564	1.564	1.536	1.825	−1.055	1.707	0.123
1.220	2.324	0.613	1.080	1.320	1.237	−2.213	−0.874	−2.041	−0.081
−0.941	0.719	0.002	−1.744	0.251	1.070	0.460	−1.598	0.656	−0.975
0.864	0.580	0.631	1.159	−0.982	−0.356	−0.023	−1.385	−2.075	+1.335

Section 3

Financial tables

Notes on financial tables

Table 29 *Amount of 1 at compound interest.* Let a given principal P (the present value of an amount of money) be invested at the rate of interest i (per period) and denote by S the compound amount of P at the end of n interest periods. Since P earns Pi interest over the first interest period, it amounts to $P + Pi = P(1 + i)$ at the end of that period. Similarly, at the end of the second interest period, the amount is $P(1 + i)(1 + i) = P(1 + i)^2$; at the end of the third interest period, the amount is $P(1 + i)^2(1 + i) = P(1 + i)^3$; and so on. These successive amounts form a geometric progression whose nth term is $S = P(1 + i)^n$.

Table 30 *Present value of 1 at compound interest.* Finding present values (or discounting as it is commonly called) is the reverse of compounding. The formula $S = P(1 + i)^n$, can be transformed into a present value formula. Dividing both sides by $(1 + i)^n$ we have:

$$P = \frac{S}{(1 + i)^n} = S(1 + i)^{-n}$$

Table 31 *Amount of annuity of 1 per period.* The amount of an annuity is defined as the value at the end of a specified number of periods of equal periodic (usually annual) payments (paid at the end of each period) and compounded at a fixed rate of interest per period. This is called an *ordinary* or *regular* annuity. Had the payment been made at the beginning of each period, the annuity would have been called an annuity *due* and each payment would have been shifted back one year. Table 31 refers to an *ordinary* annuity.

The annuity formula can be expressed algebraically, with S_n defined as the compound sum, R as the periodic payment, and n as the length of the annuity. Thus:

$$\begin{aligned} S_n &= R(1 + i)^{n-1} + R(1 + i)^{n-2} + \cdots + R(1 + i)^1 + R(1 + i)^0 \\ &= R[(1 + i)^{n-1} + (1 + i)^{n-2} + \cdots + (1 + i)^1 + 1] \\ &= R\left[\frac{(1 + i)^n - 1}{i}\right]. \end{aligned}$$

This formula gives the amount of an annuity just after the nth payment has been made. The interest factor (the expression in [ ]) is usually denoted by $s_{\overline{n}|i}$, and read as s subscript n at rate i.

Table 32 *Present value of annuity of 1 per period.* The present value of an annuity is obtained by calculating present value of each of the equal payments and summing them. It can also be viewed as present value of the amount to which the annuity will accumulate. Defining the present value of an annuity of n years as A_n, we can express it algebraically as follows:

$$\begin{aligned} A_n &= R(1 + i)^{-1} + R(1 + i)^{-2} + \cdots + R(1 + i)^{-n} \\ &= R[(1 + i)^{-1} + (1 + i)^{-2} + \cdots + (1 + i)^{-n}] \\ &= R\left[\frac{1 - (1 + i)^{-n}}{i}\right]. \end{aligned}$$

This formula gives the present value of an annuity one period before the first payment is made. The interest factor (the expression in [ ]) is usually denoted by $a_{\overline{n}|i}$, and read as a subscript n at rate i.

Table 33 *Periodic payment of annuity whose amount is 1.* When we solve the formulae of Tables 31 and 32 for R, i.e.,

$$S_n = R\, s_{\overline{n}|i} = R\left[\frac{(1 + i)^n - 1}{i}\right] \text{ and } A_n = R\, a_{\overline{n}|i} = R\left[\frac{1 - (1 + i)^{-n}}{i}\right],$$

we obtain $R = S_n \dfrac{1}{s_{\overline{n}|i}}$ and $R = A_n \dfrac{1}{a_{\overline{n}|i}}$

Table 33 gives the value of $\dfrac{1}{s_{\overline{n}|i}}$ for certain rates of interest and years. A separate table is not included for $\dfrac{1}{a_{\overline{n}|i}}$ since $\dfrac{1}{a_{\overline{n}|i}} = \dfrac{1}{s_{\overline{n}|i}} + i$. **Example:** From Table 33, $1/s_{\overline{5}|0.04} = 0.1846$ and, therefore, $1/a_{\overline{5}|0.04} = 1/s_{\overline{5}|0.04} + 0.04 = 0.1846 + 0.04 = 0.2246$.

Table 29. Amount of 1 at compound interest

$S = (1 + i)^n$ = value of 1 after n periods at compound interest i per period. **Example:** A sum of £1,000 deposited in a savings account which pays 4% interest compounded annually will grow in 5 years' time to:

$1000S = 1000(1.04)^5 = 1000(1.2167) = £1,216.70$.

n \ i	$\frac{1}{4}$%	$\frac{1}{2}$%	$\frac{2}{3}$%	$\frac{3}{4}$%	1%	$1\frac{1}{2}$%	2%	$2\frac{1}{2}$%	3%	4%	5%	6%
1	1.0025	1.0050	1.0067	1.0075	1.0100	1.0150	1.0200	1.0250	1.0300	1.0400	1.0500	1.0600
2	1.0050	1.0100	1.0134	1.0151	1.0201	1.0302	1.0404	1.0506	1.0609	1.0816	1.1025	1.1236
3	1.0075	1.0151	1.0201	1.0227	1.0303	1.0457	1.0612	1.0769	1.0927	1.1249	1.1576	1.1910
4	1.0100	1.0202	1.0269	1.0303	1.0406	1.0614	1.0824	1.1038	1.1255	1.1699	1.2155	1.2625
5	1.0126	1.0253	1.0338	1.0381	1.0510	1.0773	1.1041	1.1314	1.1593	1.2167	1.2763	1.3382
6	1.0151	1.0304	1.0407	1.0459	1.0615	1.0934	1.1262	1.1597	1.1941	1.2653	1.3401	1.4185
7	1.0176	1.0355	1.0476	1.0537	1.0721	1.1098	1.1487	1.1887	1.2299	1.3159	1.4071	1.5036
8	1.0202	1.0407	1.0546	1.0616	1.0829	1.1265	1.1717	1.2184	1.2668	1.3686	1.4775	1.5938
9	1.0227	1.0459	1.0616	1.0696	1.0937	1.1434	1.1951	1.2489	1.3048	1.4233	1.5513	1.6895
10	1.0253	1.0511	1.0687	1.0776	1.1046	1.1605	1.2190	1.2801	1.3439	1.4802	1.6289	1.7908
11	1.0278	1.0564	1.0758	1.0857	1.1157	1.1779	1.2434	1.3121	1.3842	1.5395	1.7103	1.8983
12	1.0304	1.0617	1.0830	1.0938	1.1268	1.1956	1.2682	1.3449	1.4258	1.6010	1.7959	2.0122
13	1.0330	1.0670	1.0902	1.1020	1.1381	1.2136	1.2936	1.3785	1.4685	1.6651	1.8856	2.1329
14	1.0356	1.0723	1.0975	1.1103	1.1495	1.2318	1.3195	1.4130	1.5126	1.7317	1.9799	2.2609
15	1.0382	1.0777	1.1048	1.1186	1.1610	1.2502	1.3459	1.4483	1.5580	1.8009	2.0789	2.3966
16	1.0408	1.0831	1.1122	1.1270	1.1726	1.2690	1.3728	1.4845	1.6047	1.8730	2.1829	2.5404
17	1.0434	1.0885	1.1196	1.1354	1.1843	1.2880	1.4002	1.5216	1.6528	1.9479	2.2920	2.6928
18	1.0460	1.0939	1.1271	1.1440	1.1961	1.3073	1.4282	1.5597	1.7024	2.0258	2.4066	2.8543
19	1.0486	1.0994	1.1346	1.1525	1.2081	1.3270	1.4568	1.5987	1.7535	2.1068	2.5270	3.0256
20	1.0512	1.1049	1.1421	1.1612	1.2202	1.3469	1.4859	1.6386	1.8061	2.1911	2.6533	3.2071
21	1.0538	1.1104	1.1497	1.1699	1.2324	1.3671	1.5157	1.6796	1.8603	2.2788	2.7860	3.3996
22	1.0565	1.1160	1.1574	1.1787	1.2447	1.3876	1.5460	1.7216	1.9161	2.3699	2.9253	3.6035
23	1.0591	1.1216	1.1651	1.1875	1.2572	1.4084	1.5769	1.7646	1.9736	2.4647	3.0715	3.8197
24	1.0618	1.1272	1.1729	1.1964	1.2697	1.4295	1.6084	1.8087	2.0328	2.5633	3.2251	4.0489
25	1.0644	1.1328	1.1807	1.2054	1.2824	1.4509	1.6406	1.8539	2.0938	2.6658	3.3864	4.2919
30	1.0778	1.1614	1.2206	1.2513	1.3478	1.5631	1.8114	2.0976	2.4273	3.2434	4.3219	5.7435
35	1.0913	1.1907	1.2618	1.2989	1.4166	1.6839	1.9999	2.3732	2.8139	3.9461	5.5160	7.6861
40	1.1050	1.2208	1.3045	1.3483	1.4889	1.8140	2.2080	2.6851	3.2620	4.8010	7.0400	10.2857
45	1.1189	1.2516	1.3485	1.3997	1.5648	1.9542	2.4379	3.0379	3.7816	5.8412	8.9850	13.7646
50	1.1330	1.2832	1.3941	1.4530	1.6446	2.1052	2.6916	3.4371	4.3839	7.1067	11.4674	18.4202
75	1.2059	1.4536	1.6460	1.7514	2.1091	3.0546	4.4158	6.3722	9.1789	18.9453	38.8327	79.0569
100	1.2836	1.6467	1.9435	2.1111	2.7048	4.4320	7.2446	11.8137	19.2186	50.5049	131.5013	339.3021

Table 29 – *continued*

n \ i	7%	8%	9%	10%	11%	12%	14%	16%	18%	20%	22%	24%
1	1.0700	1.0800	1.0900	1.1000	1.1100	1.1200	1.1400	1.1600	1.1800	1.2000	1.2200	1.2400
2	1.1449	1.1664	1.1881	1.2100	1.2321	1.2544	1.2996	1.3456	1.3924	1.4400	1.4884	1.5376
3	1.2250	1.2597	1.2950	1.3310	1.3676	1.4049	1.4815	1.5609	1.6430	1.7280	1.8158	1.9066
4	1.3108	1.3605	1.4116	1.4641	1.5181	1.5735	1.6890	1.8106	1.9388	2.0736	2.2153	2.3642
5	1.4026	1.4693	1.5386	1.6105	1.6851	1.7623	1.9254	2.1003	2.2878	2.4883	2.7027	2.9316
6	1.5007	1.5869	1.6771	1.7716	1.8704	1.9738	2.1950	2.4364	2.6996	2.9860	3.2973	3.6352
7	1.6058	1.7138	1.8280	1.9487	2.0762	2.2107	2.5023	2.8262	3.1855	3.5832	4.0227	4.5077
8	1.7182	1.8509	1.9926	2.1436	2.3045	2.4760	2.8526	3.2784	3.7589	4.2998	4.9077	5.5895
9	1.8385	1.9990	2.1719	2.3579	2.5580	2.7731	3.2519	3.8030	4.4355	5.1598	5.9874	6.9310
10	1.9672	2.1589	2.3674	2.5937	2.8394	3.1058	3.7072	4.4114	5.2338	6.1917	7.3046	8.5944
11	2.1049	2.3316	2.5804	2.8531	3.1518	3.4785	4.2262	5.1173	6.1759	7.4301	8.9117	10.6571
12	2.2522	2.5182	2.8127	3.1384	3.4985	3.8960	4.8179	5.9360	7.2876	8.9161	10.8722	13.2148
13	2.4098	2.7196	3.0658	3.4523	3.8833	4.3635	5.4924	6.8858	8.5994	10.6993	13.2641	16.3863
14	2.5785	2.9372	3.3417	3.7975	4.3104	4.8871	6.2613	7.9875	10.1472	12.8392	16.1822	20.3191
15	2.7590	3.1722	3.6425	4.1772	4.7846	5.4736	7.1379	9.2655	11.9737	15.4070	19.7423	25.1956
16	2.9522	3.4259	3.9703	4.5950	5.3109	6.1304	8.1372	10.7480	14.1290	18.4884	24.0856	31.2426
17	3.1588	3.7000	4.3276	5.0545	5.8951	6.8660	9.2765	12.4677	16.6722	22.1861	29.3844	38.7408
18	3.3799	3.9960	4.7171	5.5599	6.5436	7.6900	10.5752	14.4625	19.6733	26.6233	35.8490	48.0386
19	3.6165	4.3157	5.1417	6.1159	7.2633	8.6128	12.0557	16.7765	23.2144	31.9480	43.7358	59.5679
20	3.8697	4.6610	5.6044	6.7275	8.0623	9.6463	13.7435	19.4608	27.3930	38.3376	53.3576	73.8641
21	4.1406	5.0338	6.1088	7.4002	8.9492	10.8038	15.6676	22.5745	32.3238	46.0051	65.0963	91.5915
22	4.4304	5.4365	6.6586	8.1403	9.9336	12.1003	17.8610	26.1864	38.1421	55.2061	79.4175	113.5735
23	4.7405	5.8715	7.2579	8.9543	11.0263	13.5523	20.3616	30.3762	45.0076	66.2474	96.8894	140.8312
24	5.0724	6.3412	7.9111	9.8497	12.2392	15.1786	23.2122	35.2364	53.1090	79.4968	118.2050	174.6306
25	5.4274	6.8485	8.6231	10.8347	13.5855	17.0001	26.4619	40.8742	62.6686	95.3962	144.2101	216.5420
30	7.6123	10.0627	13.2677	17.4494	22.8923	29.9599	50.9502	85.8499	143.3706	237.3763	389.7579	634.8199
35	10.6766	14.7853	20.4140	28.1024	38.5749	52.7996	98.1002	180.3141	327.9973	590.6682	1053.4018	1861.0540
40	14.9745	21.7245	31.4094	45.2593	65.0009	93.0510	188.8835	378.7212	750.3783	1469.7716	2847.0378	5455.9126
45	21.0025	31.9204	48.3273	72.8905	109.5302	163.9876	363.6791	795.4438	1716.6839	3657.2620	7694.7122	15994.6902
50	29.4570	46.9016	74.3575	117.3909	184.5648	289.0022	700.2330	1670.7038	3927.3569	9100.4381	20796.5615	46890.4346
75	159.8760	321.2045	641.1909	1271.8954	2507.3988	4913.0558						
100	867.7163	2199.7613	5529.0408	13780.6123	34064.1753	83522.2657						

Table 30. Present value of 1 at compound interest

$V^n = (1 + i)^{-n}$ = present value of 1 payable at the end of period n and discounted at compound interest i per period. Example: Present value of £1,216.70 payable in 5 years' time discounted at 4% per annum is:

$1216.70(1.04)^{-5} = 1216.70(0.8219) = £1,000.$

n \ i	$\frac{1}{4}$%	$\frac{1}{2}$%	$\frac{2}{3}$%	$\frac{3}{4}$%	1%	$1\frac{1}{2}$%	2%	$2\frac{1}{2}$%	3%	4%	5%	6%
1	.9975	.9950	.9934	.9926	.9901	.9852	.9804	.9756	.9709	.9615	.9524	.9434
2	.9950	.9901	.9868	.9852	.9803	.9707	.9612	.9518	.9426	.9246	.9070	.8900
3	.9925	.9851	.9803	.9778	.9706	.9563	.9423	.9286	.9151	.8890	.8638	.8396
4	.9901	.9802	.9738	.9706	.9610	.9422	.9238	.9060	.8885	.8548	.8227	.7921
5	.9876	.9754	.9673	.9633	.9515	.9283	.9057	.8839	.8626	.8219	.7835	.7473
6	.9851	.9705	.9609	.9562	.9420	.9145	.8880	.8623	.8375	.7903	.7462	.7050
7	.9827	.9657	.9546	.9490	.9327	.9010	.8706	.8413	.8131	.7599	.7107	.6651
8	.9802	.9609	.9482	.9420	.9235	.8877	.8535	.8207	.7894	.7307	.6768	.6274
9	.9778	.9561	.9419	.9350	.9143	.8746	.8368	.8007	.7664	.7026	.6446	.5919
10	.9753	.9513	.9357	.9280	.9053	.8617	.8203	.7812	.7441	.6756	.6139	.5584
11	.9729	.9466	.9295	.9211	.8963	.8489	.8043	.7621	.7224	.6496	.5847	.5268
12	.9705	.9419	.9234	.9142	.8874	.8364	.7885	.7436	.7014	.6246	.5568	.4970
13	.9681	.9372	.9172	.9074	.8787	.8240	.7730	.7254	.6810	.6006	.5303	.4688
14	.9656	.9326	.9112	.9007	.8700	.8118	.7579	.7077	.6611	.5775	.5051	.4423
15	.9632	.9279	.9051	.8940	.8613	.7999	.7430	.6905	.6419	.5553	.4810	.4173
16	.9608	.9233	.8991	.8873	.8528	.7880	.7284	.6736	.6232	.5339	.4581	.3936
17	.9584	.9187	.8932	.8807	.8444	.7764	.7142	.6572	.6050	.5134	.4363	.3714
18	.9561	.9141	.8873	.8742	.8360	.7649	.7002	.6412	.5874	.4936	.4155	.3503
19	.9537	.9096	.8814	.8676	.8277	.7536	.6864	.6255	.5703	.4746	.3957	.3305
20	.9513	.9051	.8756	.8612	.8195	.7425	.6730	.6103	.5537	.4564	.3769	.3118
21	.9489	.9006	.8698	.8548	.8114	.7315	.6598	.5954	.5375	.4388	.3589	.2942
22	.9466	.8961	.8640	.8484	.8034	.7207	.6468	.5809	.5219	.4220	.3418	.2775
23	.9442	.8916	.8583	.8421	.7954	.7100	.6342	.5667	.5067	.4057	.3256	.2618
24	.9418	.8872	.8526	.8358	.7876	.6995	.6217	.5529	.4919	.3901	.3101	.2470
25	.9395	.8828	.8469	.8296	.7798	.6892	.6095	.5394	.4776	.3751	.2953	.2330
30	.9278	.8610	.8193	.7992	.7419	.6398	.5521	.4767	.4120	.3083	.2314	.1741
35	.9163	.8398	.7925	.7699	.7059	.5939	.5000	.4214	.3554	.2534	.1813	.1301
40	.9050	.8191	.7666	.7416	.6717	.5513	.4529	.3724	.3066	.2083	.1420	.0972
45	.8937	.7990	.7415	.7145	.6391	.5117	.4102	.3292	.2644	.1712	.1113	.0727
50	.8826	.7793	.7173	.6883	.6080	.4750	.3715	.2909	.2281	.1407	.0872	.0543
75	.8292	.6879	.6075	.5710	.4741	.3274	.2265	.1569	.1089	.0528	.0258	.0126
100	.7790	.6073	.5145	.4737	.3697	.2256	.1380	.0846	.0520	.0198	.0076	.0029

Table 30 – *continued*

n \ i	7%	8%	9%	10%	11%	12%	14%	16%	18%	20%	22%	24%
1	.9346	.9259	.9174	.9091	.9009	.8929	.8772	.8621	.8475	.8333	.8197	.8065
2	.8734	.8573	.8417	.8264	.8116	.7972	.7695	.7432	.7182	.6944	.6719	.6504
3	.8163	.7938	.7722	.7513	.7312	.7118	.6750	.6407	.6086	.5787	.5507	.5245
4	.7629	.7350	.7084	.6830	.6587	.6355	.5921	.5523	.5158	.4823	.4514	.4230
5	.7130	.6806	.6499	.6209	.5935	.5674	.5194	.4761	.4371	.4019	.3700	.3411
6	.6663	.6302	.5963	.5645	.5346	.5066	.4556	.4104	.3704	.3349	.3033	.2751
7	.6227	.5835	.5470	.5132	.4817	.4523	.3996	.3538	.3139	.2791	.2486	.2218
8	.5820	.5403	.5019	.4665	.4339	.4039	.3506	.3050	.2660	.2326	.2038	.1789
9	.5439	.5002	.4604	.4241	.3909	.3606	.3075	.2630	.2255	.1938	.1670	.1443
10	.5083	.4632	.4224	.3855	.3522	.3220	.2697	.2267	.1911	.1615	.1369	.1164
11	.4751	.4289	.3875	.3505	.3173	.2875	.2366	.1954	.1619	.1346	.1122	.0938
12	.4440	.3971	.3555	.3186	.2858	.2567	.2076	.1685	.1372	.1122	.0920	.0757
13	.4150	.3677	.3262	.2897	.2575	.2292	.1821	.1452	.1163	.0935	.0754	.0610
14	.3878	.3405	.2992	.2633	.2320	.2046	.1597	.1252	.0985	.0779	.0618	.0492
15	.3624	.3152	.2745	.2394	.2090	.1827	.1401	.1079	.0835	.0649	.0507	.0397
16	.3387	.2919	.2519	.2176	.1883	.1631	.1229	.0930	.0708	.0541	.0415	.0320
17	.3166	.2703	.2311	.1978	.1696	.1456	.1078	.0802	.0600	.0451	.0340	.0258
18	.2959	.2502	.2120	.1799	.1528	.1300	.0946	.0691	.0508	.0376	.0279	.0208
19	.2765	.2317	.1945	.1635	.1377	.1161	.0829	.0596	.0431	.0313	.0229	.0168
20	.2584	.2145	.1784	.1486	.1240	.1037	.0728	.0514	.0365	.0261	.0187	.0135
21	.2415	.1987	.1637	.1351	.1117	.0926	.0638	.0443	.0309	.0217	.0154	.0109
22	.2257	.1839	.1502	.1228	.1007	.0826	.0560	.0382	.0262	.0181	.0126	.0088
23	.2109	.1703	.1378	.1117	.0907	.0738	.0491	.0329	.0222	.0151	.0103	.0071
24	.1971	.1577	.1264	.1015	.0817	.0659	.0431	.0284	.0188	.0126	.0085	.0057
25	.1842	.1460	.1160	.0923	.0736	.0588	.0378	.0245	.0160	.0105	.0069	.0046
30	.1314	.0994	.0754	.0573	.0437	.0334	.0196	.0116	.0070	.0042	.0026	.0016
35	.0937	.0676	.0490	.0356	.0259	.0189	.0102	.0055	.0030	.0017	.0009	.0005
40	.0668	.0460	.0318	.0221	.0154	.0107	.0053	.0026	.0013	.0007	.0004	.0002
45	.0476	.0313	.0207	.0137	.0091	.0061	.0027	.0013	.0006	.0003	.0001	.0001
50	.0339	.0213	.0134	.0085	.0054	.0035	.0014	.0006	.0003	.0001	.0000	.0000
75	.0063	.0031	.0016	.0008	.0004	.0002	.0001	.0000	.0000	.0000	.0000	.0000
100	.0012	.0005	.0002	.0001	.0000	.0000	.0000	.0000	.0000	.0000	.0000	.0000

$s_{\overline{n}|i} = \dfrac{(1+i)^n - 1}{i}$ = amount of an *ordinary* annuity of 1 per period, after n periods, compounded at the rate of interest i per period. **Example:** If £1,000 is invested at the end of each year for 5 years at compound interest of 4% per annum, the value of the annuity at the end of the 5th year will be:

$$1000 s_{\overline{5}|0.04} = 1000\left(\frac{1.04^5 - 1}{0.04}\right) = 1000(5.4163) = £5,416.30.$$

n \ i	$\frac{1}{4}$%	$\frac{1}{2}$%	$\frac{2}{3}$%	$\frac{3}{4}$%	1%	$1\frac{1}{2}$%	2%	$2\frac{1}{2}$%	3%	4%	5%	6%	7%
1	1.0000	1.0000	1.0000	1.0000	1.0000	1.0000	1.0000	1.0000	1.0000	1.0000	1.0000	1.0000	1.0000
2	2.0025	2.0050	2.0067	2.0075	2.0100	2.0150	2.0200	2.0250	2.0300	2.0400	2.0500	2.0600	2.0700
3	3.0075	3.0150	3.0200	3.0226	3.0301	3.0452	3.0604	3.0756	3.0909	3.1216	3.1525	3.1836	3.2149
4	4.0150	4.0301	4.0402	4.0452	4.0604	4.0909	4.1216	4.1525	4.1836	4.2465	4.3101	4.3746	4.4399
5	5.0251	5.0503	5.0671	5.0756	5.1010	5.1523	5.2040	5.2563	5.3091	5.4163	5.5256	5.6371	5.7507
6	6.0376	6.0755	6.1009	6.1136	6.1520	6.2296	6.3081	6.3877	6.4684	6.6330	6.8019	6.9753	7.1533
7	7.0527	7.1059	7.1416	7.1595	7.2135	7.3230	7.4343	7.5474	7.6625	7.8983	8.1420	8.3938	8.6540
8	8.0704	8.1414	8.1892	8.2132	8.2857	8.4328	8.5830	8.7361	8.8923	9.2142	9.5491	9.8975	10.2598
9	9.0905	9.1821	9.2438	9.2748	9.3685	9.5593	9.7546	9.9545	10.1591	10.5828	11.0266	11.4913	11.9780
10	10.1133	10.2280	10.3054	10.3443	10.4622	10.7027	10.9497	11.2034	11.4639	12.0061	12.5779	13.1808	13.8164
11	11.1385	11.2792	11.3741	11.4219	11.5668	11.8633	12.1687	12.4835	12.8078	13.4864	14.2068	14.9716	15.7836
12	12.1664	12.3356	12.4499	12.5076	12.6825	13.0412	13.4121	13.7956	14.1920	15.0258	15.9171	16.8699	17.8885
13	13.1968	13.3972	13.5330	13.6014	13.8093	14.2368	14.6803	15.1404	15.6178	16.6268	17.7130	18.8821	20.1406
14	14.2298	14.4642	14.6232	14.7034	14.9474	15.4504	15.9739	16.5190	17.0863	18.2919	19.5986	21.0151	22.5505
15	15.2654	15.5365	15.7207	15.8137	16.0969	16.6821	17.2934	17.9319	18.5989	20.0236	21.5786	23.2760	25.1290
16	16.3035	16.6142	16.8255	16.9323	17.2579	17.9324	18.6393	19.3802	20.1569	21.8245	23.6575	25.6725	27.8881
17	17.3443	17.6973	17.9377	18.0593	18.4304	19.2014	20.0121	20.8647	21.7616	23.6975	25.8404	28.2129	30.8402
18	18.3876	18.7858	19.0572	19.1947	19.6147	20.4894	21.4123	22.3863	23.4144	25.6454	28.1324	30.9057	33.9990
19	19.4336	19.8797	20.1843	20.3387	20.8109	21.7967	22.8406	23.9460	25.1169	27.6712	30.5390	33.7600	37.3790
20	20.4822	20.9791	21.3189	21.4912	22.0190	23.1237	24.2974	25.5447	26.8704	29.7781	33.0660	36.7856	40.9955
21	21.5334	22.0840	22.4610	22.6524	23.2392	24.4705	25.7833	27.1833	28.6765	31.9692	35.7193	39.9927	44.8652
22	22.5872	23.1944	23.6107	23.8223	24.4716	25.8376	27.2990	28.8629	30.5368	34.2480	38.5052	43.3923	49.0057
23	23.6437	24.3104	24.7682	25.0010	25.7163	27.2251	28.8450	30.5844	32.4529	36.6179	41.4305	46.9958	53.4361
24	24.7028	25.4320	25.9333	26.1885	26.9735	28.6335	30.4219	32.3490	34.4265	39.0826	44.5020	50.8156	58.1767
25	25.7646	26.5591	27.1062	27.3849	28.2432	30.0630	32.0303	34.1578	36.4593	41.6459	47.7271	54.8645	63.2490
30	31.1133	32.2800	33.0890	33.5029	34.7849	37.5387	40.5681	43.9027	47.5754	56.0849	66.4388	79.0582	94.4608
35	36.5292	38.1454	39.2740	39.8538	41.6603	45.5921	49.9945	54.9282	60.4621	73.6522	90.3203	111.4348	138.2369
40	42.0132	44.1588	45.6678	46.4465	48.8864	54.2679	60.4020	67.4026	75.4013	95.0255	120.7998	154.7620	199.6351
45	47.5661	50.3242	52.2777	53.2901	56.4811	63.6142	71.8927	81.5161	92.7199	121.0294	159.7002	212.7435	285.7493
50	53.1887	56.6452	59.1109	60.3943	64.4632	73.6828	84.5794	97.4843	112.7969	152.6671	209.3480	290.3359	406.5289
75	82.3792	90.7265	96.8995	100.1833	110.9128	136.9728	170.7918	214.8883	272.6309	448.6314	756.6537	1300.9487	2269.6574
100	113.4500	129.3337	141.5170	148.1445	170.4814	228.8030	312.2323	432.5487	607.2877	1237.6237	2610.0252	5638.3681	12381.6618

Table 31 – *continued*

n \ i	8%	9%	10%	11%	12%	14%	16%	18%	20%	22%	24%
1	1.0000	1.0000	1.0000	1.0000	1.0000	1.0000	1.0000	1.0000	1.0000	1.0000	1.0000
2	2.0800	2.0900	2.1000	2.1100	2.1200	2.1400	2.1600	2.1800	2.2000	2.2200	2.2400
3	3.2464	3.2781	3.3100	3.3421	3.3744	3.4396	3.5056	3.5724	3.6400	3.7084	3.7776
4	4.5061	4.5731	4.6410	4.7097	4.7793	4.9211	5.0665	5.2154	5.3680	5.5242	5.6842
5	5.8666	5.9847	6.1051	6.2278	6.3528	6.6101	6.8771	7.1542	7.4416	7.7396	8.0484
6	7.3359	7.5233	7.7156	7.9129	8.1152	8.5355	8.9775	9.4420	9.9299	10.4423	10.9801
7	8.9228	9.2004	9.4872	9.7833	10.0890	10.7305	11.4139	12.1415	12.9159	13.7396	14.6153
8	10.6366	11.0285	11.4359	11.8594	12.2997	13.2328	14.2401	15.3270	16.4991	17.7623	19.1229
9	12.4876	13.0210	13.5795	14.1640	14.7757	16.0853	17.5185	19.0859	20.7989	22.6700	24.7125
10	14.4866	15.1929	15.9374	16.7220	17.5487	19.3373	21.3215	23.5213	25.9587	28.6574	31.6434
11	16.6455	17.5603	18.5312	19.5614	20.6546	23.0445	25.7329	28.7551	32.1504	35.9620	40.2379
12	18.9771	20.1407	21.3843	22.7132	24.1331	27.2707	30.8502	34.9311	39.5805	44.8737	50.8950
13	21.4953	22.9534	24.5227	26.2116	28.0291	32.0887	36.7862	42.2187	48.4966	55.7459	64.1097
14	24.2149	26.0192	27.9750	30.0949	32.3926	37.5811	43.6720	50.8180	59.1959	69.0100	80.4961
15	27.1521	29.3609	31.7725	34.4054	37.2797	43.8424	51.6595	60.9653	72.0351	85.1922	100.8151
16	30.3243	33.0034	35.9497	39.1899	42.7533	50.9804	60.9250	72.9390	87.4421	104.9345	126.0108
17	33.7502	36.9737	40.5447	44.5008	48.8837	59.1176	71.6730	87.0680	105.9306	129.0201	157.2534
18	37.4502	41.3013	45.5992	50.3959	55.7497	68.3941	84.1407	103.7403	128.1167	158.4045	195.9942
19	41.4463	46.0185	51.1591	56.9395	63.4397	78.9692	98.6032	123.4135	154.7400	194.2535	244.0328
20	45.7620	51.1601	57.2750	64.2028	72.0524	91.0249	115.3797	146.6280	186.6880	237.9893	303.6006
21	50.4229	56.7645	64.0025	72.2651	81.6987	104.7684	134.8405	174.0210	225.0256	291.3469	377.4648
22	55.4568	62.8733	71.4027	81.2143	92.5026	120.4360	157.4150	206.3448	271.0307	356.4432	469.0563
23	60.8933	69.5319	79.5430	91.1479	104.6029	138.2970	183.6014	244.4868	326.2369	435.8607	582.6298
24	66.7648	76.7898	88.4973	102.1742	118.1552	158.6586	213.9776	289.4945	392.4842	532.7501	723.4610
25	73.1059	84.7009	98.3471	114.4133	133.3339	181.8708	249.2140	342.6035	471.9811	650.9551	898.0916
30	113.2832	136.3075	164.4940	199.0209	241.3327	356.7868	530.3117	790.9480	1181.8816	1767.0813	2640.9164
35	172.3168	215.7108	271.0244	341.5896	431.6635	693.5727	1120.7130	1816.6516	2948.3411	4783.6447	7750.2251
40	259.0565	337.8824	442.5926	581.8261	767.0914	1342.0251	2360.7572	4163.2130	7343.8578	12936.5353	22728.8026
45	386.5056	525.8587	718.9048	986.6386	1358.2300	2590.5648	4965.2739	9531.5771	18281.3099	34971.4191	66640.3758
50	573.7702	815.0836	1163.9085	1668.7712	2400.0182	4994.5213	10435.6488	21813.0937	45497.1907	94525.2793	195372.6443
75	4002.5566	7113.2321	12708.9537	22785.4434	40933.7987						
100	27484.5157	61422.6754	137796.1235	309665.2298	696010.5474						

Table 32. Present value of annuity of 1 per period

$a_{\overline{n}|i} = \dfrac{1-(1+i)^{-n}}{i}$ = present value of an ordinary annuity of 1 per period, after n periods, discounted at compound rate of interest of i per period. **Example:** Present value of 5 annual payments of £1,000 each (paid at the end of each year) and discounted by the compound rate of interest of 4% per annum is:

$$1000a_{\overline{5}|0.04} = 1000\left(\frac{1-1.04^{-5}}{0.04}\right) = 1000(4.4518) = £4,451.80.$$

n \ i	$\frac{1}{4}$%	$\frac{1}{2}$%	$\frac{2}{3}$%	$\frac{3}{4}$%	1%	$1\frac{1}{2}$%	2%	$2\frac{1}{2}$%	3%	4%	5%	6%
1	0.9975	0.9950	0.9934	0.9926	0.9901	0.9852	0.9804	0.9756	0.9709	0.9615	0.9524	0.9434
2	1.9925	1.9851	1.9802	1.9777	1.9704	1.9559	1.9416	1.9274	1.9135	1.8861	1.8594	1.8334
3	2.9851	2.9702	2.9604	2.9556	2.9410	2.9122	2.8839	2.8560	2.8286	2.7751	2.7232	2.6730
4	3.9751	3.9505	3.9342	3.9261	3.9020	3.8544	3.8077	3.7620	3.7171	3.6299	3.5460	3.4651
5	4.9627	4.9259	4.9015	4.8894	4.8534	4.7826	4.7135	4.6458	4.5797	4.4518	4.3295	4.2124
6	5.9478	5.8964	5.8624	5.8456	5.7955	5.6972	5.6014	5.5081	5.4172	5.2421	5.0757	4.9173
7	6.9305	6.8621	6.8170	6.7946	6.7282	6.5982	6.4720	6.3494	6.2303	6.0021	5.7864	5.5824
8	7.9107	7.8230	7.7652	7.7366	7.6517	7.4859	7.3255	7.1701	7.0197	6.7327	6.4632	6.2098
9	8.8885	8.7791	8.7072	8.6716	8.5660	8.3605	8.1622	7.9709	7.7861	7.4353	7.1078	6.8017
10	9.8639	9.7304	9.6429	9.5996	9.4713	9.2222	8.9826	8.7521	8.5302	8.1109	7.7217	7.3601
11	10.8368	10.6770	10.5724	10.5207	10.3676	10.0711	9.7868	9.5142	9.2526	9.7605	8.3064	7.8869
12	11.8073	11.6189	11.4958	11.4349	11.2551	10.9075	10.5753	10.2578	9.9540	9.3851	8.8633	8.3838
13	12.7753	12.5562	12.4130	12.3423	12.1337	11.7315	11.3484	10.9832	10.6350	9.9856	9.3936	8.8527
14	13.7410	13.4887	13.3242	13.2430	13.0037	12.5434	12.1062	11.6909	11.2961	10.5631	9.8986	9.2950
15	14.7042	14.4166	14.2293	14.1370	13.8651	13.3432	12.8493	12.3814	11.9379	11.1184	10.3797	9.7122
16	15.6650	15.3399	15.1284	15.0243	14.7179	14.1313	13.5777	13.0550	12.5611	11.6523	10.8378	10.1059
17	16.6235	16.2586	16.0216	15.9050	15.5623	14.9076	14.2919	13.7122	13.1661	12.1657	11.2741	10.4773
18	17.5795	17.1728	16.9089	16.7792	16.3983	15.6726	14.9920	14.3534	13.7535	12.6593	11.6896	10.8276
19	18.5332	18.0824	17.7903	17.6468	17.2260	16.4262	15.6785	14.9789	14.3238	13.1339	12.0853	11.1581
20	19.4845	18.9874	18.6658	18.5080	18.0456	17.1686	16.3514	15.5892	14.8775	13.5903	12.4622	11.4699
21	20.4334	19.8880	19.5356	19.3628	18.8570	17.9001	17.0112	16.1845	15.4150	14.0292	12.8212	11.7641
22	21.3800	20.7841	20.3996	20.2112	19.6604	18.6208	17.6580	16.7654	15.9369	14.4511	13.1630	12.0416
23	22.3241	21.6757	21.2579	21.0533	20.4558	19.3309	18.2922	17.3321	16.4436	14.8568	13.4886	12.3034
24	23.2660	22.5629	22.1105	21.8891	21.2434	20.0304	18.9139	17.8850	16.9355	15.2470	13.7986	12.5504
25	24.2055	23.4456	22.9574	22.7188	22.0232	20.7196	19.5235	18.4244	17.4131	15.6221	14.0939	12.7834
30	28.8679	27.7941	27.1087	26.7751	25.8077	24.0158	22.3965	20.9303	19.6004	17.2920	15.3725	13.7648
35	33.4724	32.0354	31.1244	30.6827	29.4086	27.0756	24.9986	23.1452	21.4872	18.6646	16.3742	14.4982
40	38.0199	36.1722	35.0088	34.4469	32.8347	29.9158	27.3555	25.1028	23.1148	19.7928	17.1591	15.0463
45	42.5109	40.2072	38.7663	38.0732	36.0945	32.5523	29.4902	26.8330	24.5187	20.7200	17.7741	15.4558
50	46.9462	44.1428	42.4010	41.5664	39.1961	34.9997	31.4236	28.3623	25.7298	21.4822	18.2559	15.7619
75	68.3108	62.4136	58.8686	57.2027	52.5871	44.8416	38.6771	33.7227	29.7018	23.6804	19.4850	16.4558
100	88.3825	78.5426	72.8158	70.1746	63.0289	51.6247	43.0984	36.6141	31.5989	24.5050	19.8479	16.6175

Table 32 — *continued*

n \ i	7%	8%	9%	10%	11%	12%	14%	16%	18%	20%	22%	24%
1	0.9346	0.9259	0.9174	0.9091	0.9009	0.8929	0.8772	0.8621	0.8475	0.8333	0.8197	0.8065
2	1.8080	1.7833	1.7591	1.7355	1.7125	1.6901	1.6467	1.6052	1.5656	1.5278	1.4915	1.4568
3	2.6243	2.5771	2.5313	2.4869	2.4437	2.4018	2.3216	2.2459	2.1743	2.1065	2.0422	1.9813
4	3.3872	3.3121	3.2397	3.1699	3.1024	3.0373	2.9137	2.7982	2.6901	2.5887	2.4936	2.4043
5	4.1002	3.9927	3.8897	3.7908	3.6959	3.6048	3.4331	3.2743	3.1272	2.9906	2.8636	2.7454
6	4.7665	4.6229	4.4859	4.3553	4.2305	4.1114	3.8887	3.6847	3.4976	3.3255	3.1669	3.0205
7	5.3893	5.2064	5.0330	4.8684	4.7122	4.5638	4.2883	4.0386	3.8115	3.6046	3.4155	3.2423
8	5.9713	5.7466	5.5348	5.3349	5.1461	4.9676	4.6389	4.3436	4.0776	3.8372	3.6193	3.4212
9	6.5152	6.2469	5.9952	5.7590	5.5370	5.3282	4.9464	4.6065	4.3030	4.0310	3.7863	3.5655
10	7.0236	6.7101	6.4177	6.1446	5.8892	5.6502	5.2161	4.8332	4.4941	4.1925	3.9232	3.6819
11	7.4987	7.1390	6.8052	6.4951	6.2065	5.9377	5.4527	5.0286	4.6560	4.3271	4.0354	3.7757
12	7.9427	7.5361	7.1607	6.8137	6.4924	6.1944	5.6603	5.1971	4.7932	4.4392	4.1274	3.8514
13	8.3577	7.9038	7.4869	7.1034	6.7499	6.4235	5.8424	5.3423	4.9095	4.5327	4.2028	3.9124
14	8.7455	8.2442	7.7862	7.3667	6.9819	6.6282	6.0021	5.4675	5.0081	4.6106	4.2646	3.9616
15	9.1079	8.5595	8.0607	7.6061	7.1909	6.8109	6.1422	5.5755	5.0916	4.6755	4.3152	4.0013
16	9.4466	8.8514	8.3126	7.8237	7.3792	6.9740	6.2651	5.6685	5.1624	4.7296	4.3567	4.0333
17	9.7632	9.1216	8.5436	8.0216	7.5488	7.1196	6.3729	5.7487	5.2223	4.7746	4.3908	4.0591
18	10.0591	9.3719	8.7556	8.2014	7.7016	7.2497	6.4674	5.8178	5.2732	4.8122	4.4187	4.0799
19	10.3356	9.6036	8.9501	8.3649	7.8393	7.3658	6.5504	5.8775	5.3162	4.8435	4.4415	4.0967
20	10.5940	9.8181	9.1285	8.5136	7.9633	7.4694	6.6231	5.9288	5.3527	4.8696	4.4603	4.1103
21	10.8355	10.0168	9.2922	8.6487	8.0751	7.5620	6.6870	5.9731	5.3837	4.8913	4.4756	4.1212
22	11.0612	10.2007	9.4424	8.7715	8.1757	7.6446	6.7429	6.0113	5.4099	4.9094	4.4882	4.1300
23	11.2722	10.3711	9.5802	8.8832	8.2664	7.7184	6.7921	6.0442	5.4321	4.9245	4.4985	4.1371
24	11.4693	10.5288	9.7066	8.9847	8.3481	7.7843	6.8351	6.0726	5.4509	4.9371	4.5070	4.1428
25	11.6536	10.6748	9.8226	9.0770	8.4217	7.8431	6.8729	6.0971	5.4669	4.9476	4.5139	4.1474
30	12.4090	11.2578	10.2737	9.4269	8.6938	8.0552	7.0027	6.1772	5.5168	4.9789	4.5338	4.1601
35	12.9477	11.6546	10.5668	9.6442	8.8552	8.1755	7.0700	6.2153	5.5386	4.9915	4.5411	4.1644
40	13.3317	11.9246	10.7574	9.7791	8.9511	8.2438	7.1050	6.2335	5.5482	4.9966	4.5439	4.1659
45	13.6055	12.1084	10.8812	9.8628	9.0079	8.2825	7.1232	6.2421	5.5523	4.9986	4.5449	4.1664
50	13.8007	12.2335	10.9617	9.9148	9.0417	8.3045	7.1327	6.2463	5.5541	4.9995	4.5452	4.1666
75	14.1964	12.4611	11.0938	9.9921	9.0873	8.3316	7.1425	6.2499	5.5555	5.0000	4.5455	4.1667
100	14.2693	12.4943	11.1091	9.9993	9.0906	8.3332	7.1428	6.2500	5.5556	5.0000	4.5455	4.1667

$\frac{1}{s_{\overline{n}|i}} = \frac{i}{(1+i)^n - 1}$ = periodic payment required to accumulate to an amount equal to 1 at the end of n periods when the compound rate of interest is i per period. The payments are made at the end of the periods. **Example:** The annual deposit in a savings account paying interest at 4% required to accumulate to £5,416.30 at the end of 5 years is:

$$5416.30\left(\frac{1}{s_{\overline{n}|i}}\right) = 5416.30\left(\frac{0.04}{1.04^5 - 1}\right) = 5416.30(0.1846) = £1,000.$$

n \ i	¼%	½%	⅔%	¾%	1%	1½%	2%	2½%	3%	4%	5%	6%
1	1.0000	1.0000	1.0000	1.0000	1.0000	1.0000	1.0000	1.0000	1.0000	1.0000	1.0000	1.0000
2	0.4994	0.4988	0.4983	0.4981	0.4975	0.4963	0.4950	0.4938	0.4926	0.4902	0.4878	0.4854
3	0.3325	0.3317	0.3311	0.3308	0.3300	0.3284	0.3268	0.3251	0.3235	0.3203	0.3172	0.3141
4	0.2491	0.2481	0.2475	0.2472	0.2463	0.2444	0.2426	0.2408	0.2390	0.2355	0.2320	0.2286
5	0.1990	0.1980	0.1974	0.1970	0.1960	0.1941	0.1922	0.1902	0.1884	0.1846	0.1810	0.1774
6	0.1656	0.1646	0.1639	0.1636	0.1625	0.1605	0.1585	0.1565	0.1546	0.1508	0.1470	0.1434
7	0.1418	0.1407	0.1400	0.1397	0.1386	0.1366	0.1345	0.1325	0.1305	0.1266	0.1228	0.1191
8	0.1239	0.1228	0.1221	0.1218	0.1207	0.1186	0.1165	0.1145	0.1125	0.1085	0.1047	0.1010
9	0.1100	0.1089	0.1082	0.1078	0.1067	0.1046	0.1025	0.1005	0.0984	0.0945	0.0907	0.0870
10	0.0989	0.0978	0.0970	0.0967	0.0956	0.0934	0.0913	0.0893	0.0872	0.0833	0.0795	0.0759
11	0.0898	0.0887	0.0879	0.0876	0.0865	0.0843	0.0822	0.0801	0.0781	0.0741	0.0704	0.0668
12	0.0822	0.0811	0.0803	0.0800	0.0788	0.0767	0.0746	0.0725	0.0705	0.0666	0.0628	0.0593
13	0.0758	0.0746	0.0739	0.0735	0.0724	0.0702	0.0681	0.0660	0.0640	0.0601	0.0565	0.0530
14	0.0703	0.0691	0.0684	0.0680	0.0669	0.0647	0.0626	0.0605	0.0585	0.0547	0.0510	0.0476
15	0.0655	0.0644	0.0636	0.0632	0.0621	0.0599	0.0578	0.0558	0.0538	0.0499	0.0463	0.0430
16	0.0613	0.0602	0.0594	0.0591	0.0579	0.0558	0.0537	0.0516	0.0496	0.0458	0.0423	0.0390
17	0.0577	0.0565	0.0557	0.0554	0.0543	0.0521	0.0500	0.0479	0.0460	0.0422	0.0387	0.0354
18	0.0544	0.0532	0.0525	0.0521	0.0510	0.0488	0.0467	0.0447	0.0427	0.0390	0.0355	0.0324
19	0.0515	0.0503	0.0495	0.0492	0.0481	0.0459	0.0438	0.0418	0.0398	0.0361	0.0327	0.0296
20	0.0488	0.0477	0.0469	0.0465	0.0454	0.0432	0.0412	0.0391	0.0372	0.0336	0.0302	0.0272
21	0.0464	0.0453	0.0445	0.0441	0.0430	0.0409	0.0388	0.0368	0.0349	0.0313	0.0280	0.0250
22	0.0443	0.0431	0.0424	0.0420	0.0409	0.0387	0.0366	0.0346	0.0327	0.0292	0.0260	0.0230
23	0.0423	0.0411	0.0404	0.0400	0.0389	0.0367	0.0347	0.0327	0.0308	0.0273	0.0241	0.0213
24	0.0405	0.0393	0.0386	0.0382	0.0371	0.0349	0.0329	0.0309	0.0290	0.0256	0.0225	0.0197
25	0.0388	0.0377	0.0369	0.0365	0.0354	0.0333	0.0312	0.0293	0.0274	0.0240	0.0210	0.0182
30	0.0321	0.0310	0.0302	0.0298	0.0287	0.0266	0.0246	0.0228	0.0210	0.0178	0.0151	0.0126
35	0.0274	0.0262	0.0255	0.0251	0.0240	0.0219	0.0200	0.0182	0.0165	0.0136	0.0111	0.0090
40	0.0238	0.0226	0.0219	0.0215	0.0205	0.0184	0.0166	0.0148	0.0133	0.0105	0.0083	0.0065
45	0.0210	0.0199	0.0191	0.0188	0.0177	0.0157	0.0139	0.0123	0.0108	0.0083	0.0063	0.0047
50	0.0188	0.0177	0.0169	0.0166	0.0155	0.0136	0.0118	0.0103	0.0089	0.0066	0.0048	0.0034
75	0.0121	0.0110	0.0103	0.0100	0.0090	0.0073	0.0059	0.0047	0.0037	0.0022	0.0013	0.0008
100	0.0088	0.0077	0.0071	0.0068	0.0059	0.0044	0.0032	0.0023	0.0016	0.0008	0.0004	0.0002

Table 33 – *continued*

n \ i	7%	8%	9%	10%	11%	12%	14%	16%	18%	20%	22%	24%
1	1.0000	1.0000	1.0000	1.0000	1.0000	1.0000	1.0000	1.0000	1.0000	1.0000	1.0000	1.0000
2	0.4831	0.4808	0.4785	0.4762	0.4739	0.4717	0.4673	0.4630	0.4587	0.4545	0.4505	0.4464
3	0.3111	0.3080	0.3051	0.3021	0.2992	0.2963	0.2907	0.2853	0.2799	0.2747	0.2697	0.2647
4	0.2252	0.2219	0.2187	0.2155	0.2123	0.2092	0.2032	0.1974	0.1917	0.1863	0.1810	0.1759
5	0.1739	0.1705	0.1671	0.1638	0.1606	0.1574	0.1513	0.1454	0.1398	0.1344	0.1292	0.1242
6	0.1398	0.1363	0.1329	0.1296	0.1264	0.1232	0.1172	0.1114	0.1059	0.1007	0.0958	0.0911
7	0.1156	0.1121	0.1087	0.1054	0.1022	0.0991	0.0932	0.0876	0.0824	0.0774	0.0728	0.0684
8	0.0975	0.0940	0.0907	0.0874	0.0843	0.0813	0.0756	0.0702	0.0652	0.0606	0.0563	0.0523
9	0.0835	0.0801	0.0768	0.0736	0.0706	0.0677	0.0622	0.0571	0.0524	0.0481	0.0441	0.0405
10	0.0724	0.0690	0.0658	0.0627	0.0598	0.0570	0.0517	0.0469	0.0425	0.0385	0.0349	0.0316
11	0.0634	0.0601	0.0569	0.0540	0.0511	0.0484	0.0434	0.0389	0.0348	0.0311	0.0278	0.0249
12	0.0559	0.0527	0.0497	0.0468	0.0440	0.0414	0.0367	0.0324	0.0286	0.0253	0.0223	0.0196
13	0.0497	0.0465	0.0436	0.0408	0.0382	0.0357	0.0312	0.0272	0.0237	0.0206	0.0179	0.0156
14	0.0443	0.0413	0.0384	0.0357	0.0332	0.0309	0.0266	0.0229	0.0197	0.0169	0.0145	0.0124
15	0.0398	0.0368	0.0341	0.0315	0.0291	0.0268	0.0228	0.0194	0.0164	0.0139	0.0117	0.0099
16	0.0359	0.0330	0.0303	0.0278	0.0255	0.0234	0.0196	0.0164	0.0137	0.0114	0.0095	0.0079
17	0.0324	0.0296	0.0270	0.0247	0.0225	0.0205	0.0169	0.0140	0.0115	0.0094	0.0078	0.0064
18	0.0294	0.0267	0.0242	0.0219	0.0198	0.0179	0.0146	0.0119	0.0096	0.0078	0.0063	0.0051
19	0.0268	0.0241	0.0217	0.0195	0.0176	0.0158	0.0127	0.0101	0.0081	0.0065	0.0051	0.0041
20	0.0244	0.0219	0.0195	0.0175	0.0156	0.0139	0.0110	0.0087	0.0068	0.0054	0.0042	0.0033
21	0.0223	0.0198	0.0176	0.0156	0.0138	0.0122	0.0095	0.0074	0.0057	0.0044	0.0034	0.0026
22	0.0204	0.0180	0.0159	0.0140	0.0123	0.0108	0.0083	0.0064	0.0048	0.0037	0.0028	0.0021
23	0.0187	0.0164	0.0144	0.0126	0.0110	0.0096	0.0072	0.0054	0.0041	0.0031	0.0023	0.0017
24	0.0172	0.0150	0.0130	0.0113	0.0098	0.0085	0.0063	0.0047	0.0035	0.0025	0.0019	0.0014
25	0.0158	0.0137	0.0118	0.0102	0.0087	0.0075	0.0055	0.0040	0.0029	0.0021	0.0015	0.0011
30	0.0106	0.0088	0.0073	0.0061	0.0050	0.0041	0.0028	0.0019	0.0013	0.0008	0.0006	0.0004
35	0.0072	0.0058	0.0046	0.0037	0.0029	0.0023	0.0014	0.0009	0.0006	0.0003	0.0002	0.0001
40	0.0050	0.0039	0.0030	0.0023	0.0017	0.0013	0.0007	0.0004	0.0002	0.0001	0.0001	0.0000
45	0.0035	0.0026	0.0019	0.0014	0.0010	0.0007	0.0004	0.0002	0.0001	0.0001	0.0000	0.0000
50	0.0025	0.0017	0.0012	0.0009	0.0006	0.0004	0.0002	0.0001	0.0000	0.0000	0.0000	0.0000
75	0.0004	0.0002	0.0001	0.0001	0.0000	0.0000	0.0000	0.0000	0.0000	0.0000	0.0000	0.0000
100	0.0001	0.0000	0.0000	0.0000	0.0000	0.0000	0.0000	0.0000	0.0000	0.0000	0.0000	0.0000